Palgrave Studies in Impact Finance

Series Editor
Mario La Torre, Department of Management, Sapienza
University of Rome, Rome, Italy

The *Palgrave Studies in Impact Finance* series provides a valuable scientific 'hub' for researchers, professionals and policy makers involved in Impact finance and related topics. It includes studies in the social, political, environmental and ethical impact of finance, exploring all aspects of impact finance and socially responsible investment, including policy issues, financial instruments, markets and clients, standards, regulations and financial management, with a particular focus on impact investments and microfinance.

Titles feature the most recent empirical analysis with a theoretical approach, including up to date and innovative studies that cover issues which impact finance and society globally.

Giuseppe Rimo · Simona Cosma

Biodiversity and Banks

Managing Nature Risks and Opportunities

palgrave
macmillan

Giuseppe Rimo
Department of Management
University of Bologna
Bologna, Italy

Simona Cosma
Department of Management
University of Bologna
Bologna, Italy

ISSN 2662-5105 ISSN 2662-5113 (electronic)
Palgrave Studies in Impact Finance
ISBN 978-3-032-17630-1 ISBN 978-3-032-17631-8 (eBook)
https://doi.org/10.1007/978-3-032-17631-8

This Palgrave Macmillan imprint is published by the registered company Springer Nature Switzerland AG
The registered company address is: Gewerbestrasse 11, 6330 Cham, Switzerland

Preface

Technological progress has enabled humanity to redefine possibilities, reshape our lives, expand our reach, and inspire new ways of thinking. However, behind this brilliance lies a long-ignored cost: the irreversible erosion of the natural systems that support our economies. The depletion of forests, coral reefs, species, and habitats is no longer an abstract environmental concern; it represents a systemic risk. Just as the flap of a butterfly's wings can result in the development of a tornado thousands of kilometres away, the loss of a single ecosystem can lead to a financial crisis, affecting supply chains, markets, and ultimately the global financial system. This book aims to explain why this understanding is crucial for banks.

Recently, biodiversity has entered the financial lexicon. Today, it is recognised by regulators, scholars, and firms as a material financial risk. This new awareness requires a transition from financing growth at the expense of nature to financing growth through nature. Banks can have an active role in this shift, minimising harmful capital flows and mobilising investment for restoration through innovative credit, market, and blended-finance instruments. These tools can transform biodiversity from a sidelined issue into a strategic financial variable and a new frontier for banking innovation.

The financial relevance of biodiversity loss has long been neglected but has gained increasing and urgent attention in recent years. This book explains how biodiversity loss threatens the financial health of firms that depend on nature and, in turn, the stability of the banks that fund them. Its aim is clear: to demonstrate why biodiversity loss is a financial issue and why banks must take action now. By adopting nature-positive strategies to protect natural capital, finance can become a force for resilience rather than a catalyst for collapse.

Bologna, Italy

Giuseppe Rimo
Simona Cosma

Acknowledgments The authors acknowledge the use of large language model tools (*ChatGPT 5.1*) for language refinement and drafting support in accordance with Springer Nature's AI usage policy.

Competing Interests The authors have no competing interests to declare that are relevant to the content of this manuscript.

Ethics Approval This volume does not involve studies with human participants or animals. No ethics approval was required.

Contents

About the Authors

Giuseppe Rimo is a research fellow at the Department of Management of the *Alma Mater Studiorum University of Bologna*. He gained his Ph.D. at the *University of Salento* (Italy), with a dissertation on biodiversity loss and banks' financial stability. His research interests concern biodiversity finance as well as emerging risks in the banking sector and corporate governance. He is the author and co-author of several articles published in national and international academic journals.

Simona Cosma is an associate professor of Financial Markets and Institutions at the Department of Management of the *Alma Mater Studiorum University of Bologna*. She gained her Ph.D. at the *University of Rome Tor Vergata* (Italy), with a dissertation on operational risk measurement techniques. Her research focuses on risk management in financial institutions, corporate governance, sustainability, and—most recently—biodiversity and nature-related financial risks. She is the author and co-author of numerous books and articles published in national and international academic journals.

List of Figures

List of Tables

CHAPTER 1

Introduction

Abstract This chapter begins with a brief definition of biodiversity and its significance for humans. It then discusses the accelerating loss of biodiversity, characterising it as both an ecological crisis and a source of systemic financial risk. Recent evidence suggests an unprecedented decline in species, habitats, and genetic diversity, posing a threat to the stability and resilience of natural systems. The chapter introduces the principle of double materiality, which highlights how biodiversity loss affects businesses financially (referred to as outside-in) while also noting that businesses contribute to ecological degradation (referred to as inside-out) through practices such as land use change, pollution, and unsustainable resource use. In this context, banks play a crucial role: their lending and investment decisions can either exacerbate biodiversity pressures or help mitigate them. At the same time, the decline of ecosystems can negatively impact credit risk, collateral values, supply chain stability, and overall financial performance. Regulators and standard-setters increasingly recognise the necessity of integrating nature-related risks into financial governance and supervision. The chapter concludes by outlining the structure of the volume, which offers an interdisciplinary analysis of risk channels, banking exposure, measurement tools, and the implications of biodiversity loss for financial stability and bank behaviour.

Keywords Nature · Biodiversity loss · Biodiversity finance · Financial stability · Double materiality

G. Rimo and S. Cosma, *Biodiversity and Banks*, Palgrave Studies in Impact Finance, https://doi.org/10.1007/978-3-032-17631-8_1

1.1 Background and Motivation

The term "biodiversity" refers to the variety of animals, plants, and microorganisms present on Earth as well as the ecological systems in which they live and evolve.[1] The coexistence of these varieties in a balanced condition guarantees humans everything they need to live. In fact, biodiversity gives rise to the so-called "ecosystem services", that is, all those benefits that human beings obtain from ecosystems and that are produced by interactions within the ecosystems. Examples of such services include food, drinking water, wood and minerals, protection against natural hazards, and the absorption and storage of carbon by vegetation.

During the 1980s, growing concern about the rapid decline of animal and plant species prompted research to investigate the consequences of this phenomenon. Climate change, as well as continued growth in human population, per capita consumption, and exploitation of natural resources, has, in fact, led the planet to experience a crisis characterised by the extinction of wild mammals, plants, and marine environments (Cosma et al., 2023; Dasgupta, 2021). This loss could completely undermine the structure and functioning of entire ecosystems, as well as the regulation of processes essential for providing goods and services to humanity (Cardinale, 2012; Cardinale et al., 2012). Since then, the study of the impact of biodiversity on the Earth's balance has become increasingly important. With the Rio Earth Summit in 1992, for the first time, the conservation of biodiversity became a significant issue recognised by leaders around the world.

More than two decades after the Earth Summit, biodiversity loss remains a problem that appears to have reached unprecedented levels of severity. The *Living Planet Index* (Fig. 1.1) shows that from 1970 to 2020, the size of animal populations for which data are available decreased by an average of 73% (WWF, 2024). Furthermore, even species that are not currently endangered are rapidly losing a significant portion of their genetic diversity (Karolyi & Tobin-de la Puente, 2023).

[1] https://www.isprambiente.gov.it/it/attivita/biodiversita/le-domande-piu-frequenti-sulla-biodiversita/cose-la-biodiversita.

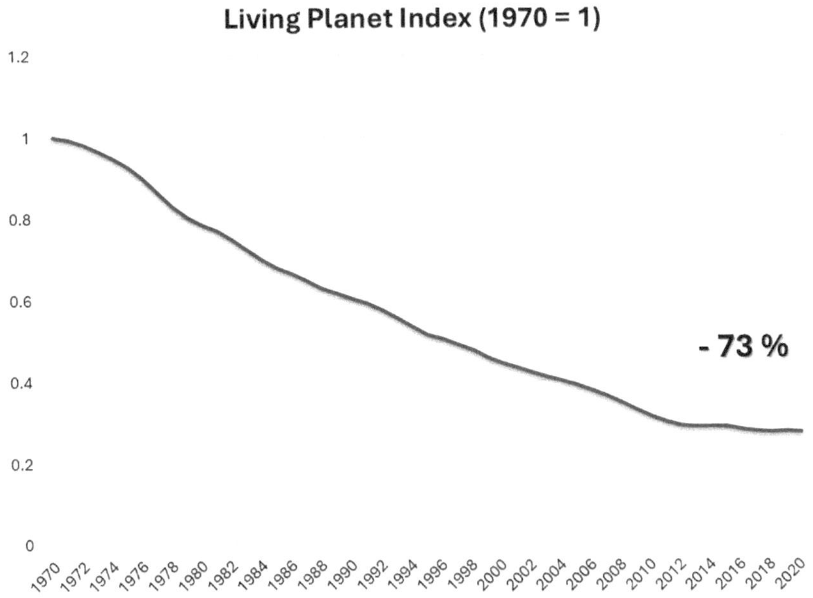

Fig. 1.1 Global living planet index. *Source* Authors' elaboration on data from 2024 Living Planet Report (WWF, 2024)

1.2 The Double Materiality of Biodiversity Loss

The strong link between biodiversity and the economy exacerbates the severity of biodiversity loss. According to an analysis by the World Economic Forum, approximately half of the world's GDP depends on nature (World Economic Forum, 2024). This figure was recently updated by an analysis published by *PwC* in April 2023, according to which 55% of the world's GDP depends on nature to a great or moderate extent.[2] In fact, biodiversity loss is not only an environmental concern but also a significant economic concern. It is a financial issue that must be understood through the lens of double materiality: economic actors are not only exposed to biodiversity loss (outside-in), but they also actively contribute to increasing it (inside-out).

[2] https://www.pwc.com/gx/en/news-room/press-releases/2023/pwcboosts-global-nature-and-biodiversity-capabilities.html.

Many industries depend, often invisibly, on the services provided by functioning ecosystems. Direct dependencies are evident in sectors such as agriculture, forestry, fisheries, and pharmaceuticals, which rely on fertile soil, pollinators, plants, genetic resources, and clean water (Katic et al., 2023; Giglio et al., 2026). However, even seemingly unrelated sectors, such as aviation, chemicals, transportation, and consumer goods, display hidden dependencies through their supply chains. These dependencies expose firms to operational and supply chain risks. As biodiversity declines, these services become less reliable or more expensive, leading to supply chain disruptions, productivity losses, and ultimately, financial instability for firms.

On the other hand, economic activities are a major driver of biodiversity loss, primarily through:

- Land use change and habitat destruction: Urban expansion, industrial agriculture, mining, and infrastructure development often lead to deforestation and the conversion of natural habitats. Corporations involved in agriculture, real estate, and extractives are directly linked to such transformations.
- Pollution: Industrial processes, agriculture, and manufacturing contribute to air, soil, and water pollution. This includes pesticide and fertiliser runoff, plastic waste, and untreated industrial discharge, all of which harm ecosystems and reduce biodiversity.
- Overexploitation and unsustainable resource use: The unsustainable extraction of raw materials, such as timber, fish, and freshwater, and the overuse of biological resources erode the regenerative capacity of ecosystems. For example, the global demand for critical minerals such as lithium and cobalt, which are essential for the green energy transition, drives mining activities that often occur in ecologically sensitive areas (Garel et al., 2024).

In this sense, economic activities are both a major driver of biodiversity loss and simultaneously highly exposed to its consequences.

In this context, financial institutions, and banks in particular, play a role of paramount importance. On the one hand, their lending and investment decisions can exacerbate biodiversity loss when capital is directed towards high-impact sectors, such as intensive agriculture, mining, fossil fuel extraction, or infrastructure projects lacking environmental safeguards. On the other side, biodiversity degradation feeds back onto banks: ecosystem decline disrupts supply chains, lowers firms' productivity, reduces collateral values, and increases default risk, thereby threatening asset quality and financial stability.

This finance-biodiversity nexus is increasingly recognised by regulators and institutions, such as the Network for Greening the Financial System (NGFS) and the Taskforce on Nature-related Financial Disclosures (TNFD), which emphasise that financial flows must be aligned with biodiversity-positive pathways.

Thus, biodiversity loss poses systemic financial risks. By fully internalising the dual materiality of biodiversity—namely, the fact that businesses and economic actors are both vulnerable to and responsible for biodiversity loss—corporate and financial decision-makers can better integrate nature-related risks into their governance, risk management, and capital allocation frameworks. Biodiversity is not just a sustainability issue: it is a driver of long-term economic resilience.

1.3 Volume Overview

This volume aims to demonstrate the urgent need to integrate biodiversity loss into the financial risk agenda, thereby bridging a critical gap in both academic literature and regulatory practice. Its originality lies in framing biodiversity not only as an ecological concern but as a systemic financial risk, with relevant and tangible implications for banking stability, credit allocation, and regulatory innovation. The structure of the volume reflects these aims. Chapter 2 analyses the channels through which biodiversity loss affects the financial system. Chapter 3 provides a systematic review and bibliometric mapping of the rapidly expanding literature on biodiversity and banking. Chapter 4 discusses how biodiversity-related impacts and dependencies can be measured through sectoral, geographical, and firm-level tools. Chapter 5 turns to the financing side and examines the expanding landscape of biodiversity-related financial instruments. Chapter 6 presents original empirical evidence on how banks integrate biodiversity and environmental risk considerations into financing decisions, and whether banks' attention to nature-related risks is associated with differences in performance, risk-taking, and resilience.

Together, these chapters offer a novel and interdisciplinary framework that combines theoretical foundations, empirical evidence, and policy-relevant analysis to guide scholars, regulators, and practitioners in rethinking banking in the age of nature risk.

References

Cardinale, B. (2012). Impacts of biodiversity loss. *Science, 336*(6081), 552–553.

Cardinale, B. J., Duffy, J. E., Gonzalez, A., Hooper, D. U., Perrings, C., Venail, P., & Naeem, S. (2012). Biodiversity loss and its impact on humanity. *Nature, 486*(7401), 59–67.

Cosma, S., Rimo, G., & Cosma, S. (2023). Conservation finance: What are we not doing? A review and research agenda. *Journal of Environmental Management, 336*, Article 117649.

Dasgupta, P. (2021). The Economics of Biodiversity: The Dasgupta. London: HM Treasury.

Garel, A., Romec, A., Sautner, Z., & Wagner, A. F. (2024). Do investors care about biodiversity? *Review of Finance, 28*(4), 1151–1186.

Giglio, S., Kuchler, T., Stroebel, J., & Zeng, X. (2026). Biodiversity risk. *Review of Finance, 30*(1), 131–161.

Karolyi, G. A., & Tobin-de la Puente, J. (2023). Biodiversity finance: A call for research into financing nature. *Financial Management, 52*(2), 231–251.

Katic, P. G., Cerretelli, S., Haggar, J., Santika, T., & Walsh, C. (2023). Mainstreaming biodiversity in business decisions: Taking stock of tools and gaps. *Biological Conservation, 277*, Article 109831.

WWF. (2024). *Living planet report 2024—A system in peril*. WWF.

CHAPTER 2

Biodiversity-Related Risks: Impacts, Dependencies and Regulation

Abstract This chapter examines how nature degradation can affect financial stability, with a focus on the Eurozone. Through real-world examples and modelling work from the European Central Bank, we show that financial institutions are already exposed to the economic consequences of the decline of ecosystems. However, finance is not neutral: it directs capital flows, and it can either fuel biodiversity loss or help to halt it. That is why financial institutions need to adopt a dual role: they must enhance their assessment and management of nature-related risks, while also supporting the transition to activities that protect and restore biodiversity. We draw on economic principles and recent EU regulatory steps, including the Taxonomy, CSRD, SFDR, and ECB/EBA guidance, to show how nature is being brought into the financial rulebook. What is needed now is not just environmental awareness, but a more profound rethink of how finance works, so it can manage risks more wisely and help build a more sustainable future.

Keywords Biodiversity loss · Biodiversity risk · Physical risk · Transition risk · Financial institutions

G. Rimo and S. Cosma, *Biodiversity and Banks*, Palgrave Studies in Impact Finance, https://doi.org/10.1007/978-3-032-17631-8_2

2.1 Biodiversity Loss and Financial Risks

The Global Risks Report 2025, published by the World Economic Forum, identifies biodiversity loss and ecosystem collapse as one of the most severe long-term global risks, ranking second in perceived impact over a 10-year horizon. While biodiversity risk is not among the top concerns for the next 2 years, it shows one of the sharpest increases in severity between short- and long-term projections (WEF, 2025).

This escalation reflects a growing awareness of the systemic dependencies of economies, food systems, health, and climate resilience on the functioning of ecosystems. As the report emphasises, the deterioration of biodiversity is not just an environmental issue but also a macroeconomic and financial risk. In a typical scenario, the costs of biodiversity loss in some countries could reach 4% of their annual GDP by 2050 (Johnson et al., 2020). The progressive degradation of nature can compromise production processes and weaken the solvency of numerous industrial sectors, resulting in *physical and transition risks*.

Physical risks associated with biodiversity loss manifest through both acute and chronic risks. *Acute risks* include increasingly intense and frequent natural disasters, such as floods, wildfires, and the spread of pests that destroy crops due to the disappearance of natural predators, exacerbated by degraded ecosystems that can no longer buffer shocks. These shock-type events occur rapidly, can cause severe damage, and are difficult to predict. *Chronic risks*, on the other hand, stem from the gradual erosion of ecosystems' capacity to function, for example, through soil degradation, coral bleaching, and deforestation (Boldrini et al., 2023). They have long-term, gradual impacts that unfold over time as ecosystems degrade and erode the resilience of companies and communities slowly but steadily.

Both the welfare of people and economic systems are impacted by these changes. For example, farming and food security are at risk due to the decline of pollinator species, such as bees and butterflies. Freshwater supplies and natural flood control are diminished by deforestation and wetland loss. For both urban and rural populations, this raises the risk of water shortages and damage from extreme events.

An acute biodiversity risk was clearly visible in the summer of 2022 when shellfish farmers in Brittany and Normandy faced a crisis. The sea was contaminated by a massive algal bloom triggered by warmer waters and an excess of nutrients from sewage and agricultural runoff. Oysters and mussels accumulated these toxins, rendering them unfit for human consumption. Harvesting ceased overnight, sales plummeted, and small producers lost as much as 80% of their seasonal production.

An example of chronic biodiversity risk is evident in recent years in the United Kingdom. Farmers are struggling with the steady decline of wild pollinators, including bees, butterflies, and hoverflies. This is not caused by a single event but by decades of intensive farming, pesticide use, and habitat loss. For fruit and nut growers, the impact is significant: lower yields, higher costs to hire managed bees or plant wildflowers, and shrinking margins, not from sudden shocks, but from the gradual decline of a vital ecosystem service.

The decline in biodiversity is also linked to climate change. For instance, it reduces the ability of land and ocean ecosystems to capture carbon. This weakens efforts to fight climate change. Resource scarcity also has sector-specific implications: for example, shortages of bioactive compounds derived from rare plants or microorganisms can disrupt pharmaceutical supply chains, while degraded landscapes and the disappearance of wildlife make many natural tourism destinations less appealing, affecting local economies that depend on ecotourism.

Transition risks affect economic sectors that impact nature or are heavily dependent on it (and are often among those that damage it most). Companies belonging to these sectors may face significant reputational and adaptation risks to new environmental protection regulations, to the new sensitivity and attention of demand to the issue of biodiversity loss and to the physiological decay or change in quantity and quality of natural resources that are the primary inputs of the production process (Rimo et al., 2024). Governments and international bodies have tried to conserve and restore natural capital in recent years. A key milestone in this direction is the Kunming-Montreal Global Biodiversity Framework, adopted at the 15th Conference of the Parties (COP15) to the UN Convention on Biological Diversity in December 2022. This framework sets ambitious global targets, such as protecting 30% of land and marine areas by 2030, and provides a basis for national policies that may reshape entire sectors.

These policy changes may include limits on raw material extraction, bans on products linked to deforestation or habitat destruction, mandatory disclosures about the impact on nature, or pricing methods that account for biodiversity impacts. Implementing these policies poses regulatory risks for businesses that harm the environment or rely on ecosystem services. Non-financial companies in agriculture, mining, fashion, and tourism may need to change their supply chains, invest in practices that support biodiversity, or eliminate entire product lines. As a result, business models that do not fit with new environmental standards and societal expectations may become economically unviable. Stranded assets, restructuring costs, or loss of competitive advantage are examples of transition costs, particularly for businesses that put off change.

For instance, a new law aimed at addressing imported deforestation was passed by the European Union in 2023 (*EU Deforestation Regulation*). This rule prohibits businesses from selling EU goods linked to ecosystem harm or forest loss. It impacts key commodities such as cocoa, coffee, palm oil, soy, beef, and timber, all of which are closely tied to biodiversity loss. One of the world's largest food corporations, Nestlé, relies heavily on several key ingredients, particularly cocoa and palm oil. To remain compliant, the business had to make significant, structural adjustments to its supply chains. To meet the new regulatory requirements, Nestlé shifted from traditional supplier audits to advanced traceability systems, including satellite monitoring and geo-tagging, substantially increasing visibility across its cocoa and palm-oil sourcing networks.[1] It severed ties with suppliers in Southeast Asia and West Africa that were unable to demonstrate that their products originated from deforestation-free areas. Additionally, Nestlé began employing more robust "zero-deforestation" certification systems and increased its investments in monitoring the effects on biodiversity. These changes were not inexpensive. They did, however, assist Nestlé in preserving its access to the European market, avoiding potential penalties, and establishing its reputation with investors who value sustainability.

Aside from regulations, new technology, changes in consumer preferences, and pressure from investors are also speeding up the transition. Companies that fail to meet these new expectations may face higher capital costs, damage to their reputation, or limited market access. The abovementioned regulation has generated a transition risk for Lavazza, which publicly requested a further postponement of its entry into force, arguing that the coffee industry is not yet ready to comply with the new

[1] https://www.foodnavigator.com/Article/2024/06/25/How-is-Nestle-making-cocoa-more-traceable/.

traceability and due diligence requirements.[2] As a result, the company has faced reputational risks (public petitions, media coverage, and NGO campaigns linking the brand to deforestation and biodiversity loss), a loss of consumer trust, and a potential legal risk if its environmental claims are considered misleading (greenwashing) in light of its actual practices.

On the other hand, early adopters of nature-positive innovation may gain from increased resilience and new market opportunities; however, they can also expose themselves to reputational, social, and market risks. Ecover, the Belgian company known for its eco-friendly cleaning products, was at the centre of controversy when it tried to innovate. To reduce its dependence on palm oil, the company introduced a new ingredient derived from synthetic algae oil.[3] This product was presented as a more sustainable option. Consumer groups and environmental organisations, such as the Naturewatch Foundation, expressed concerns. They questioned the use of genetically engineered algae, the lack of transparency in sourcing, and the broader ecological risks associated with synthetic biology.[4] Ecover lost key ethical certifications, including its endorsement in the Compassionate Shopping Guide. Environmental campaigners urged consumers to boycott the brand. The company faced growing reputational pressure, with increased scrutiny over its marketing claims and sustainability credentials. While Ecover did not go bankrupt, the episode was a clear warning. It suffered reputational damage, likely lost customer trust, and had to manage the costs of public criticism and brand repair, all triggered by a shift in consumer expectations around biodiversity and the ethics of "natural" products.

[2] https://www.reuters.com/sustainability/climate-energy/coffee-maker-lavazza-calls-further-delay-eu-deforestation-law-2025-07-09/.

[3] https://www.theguardian.com/environment/2014/apr/02/ecover-algae-laundry-liquid-palm-oil.

[4] https://theecologist.org/2014/jun/16/ecover-pioneers-synthetic-biology-consumer-products

2.2 Economy and Biodiversity: A Strict Impact-Dependence Relation

The concerns described above regarding the financial risks of biodiversity loss are confirmed by empirical analyses and structured assessments from authoritative institutions, which have begun to systematically quantify both the dependencies of economic sectors on ecosystem services and the adverse impacts that corporate and sectoral activities exert on biodiversity.

According to the Finance for Biodiversity Foundation's, 2024 report, the economic activities of publicly listed companies have a significant impact on biodiversity. A small number of firms and sectors contribute a large portion of the estimated global impact. In particular, the food, oil and gas, and chemicals industries are the most significant contributors to biodiversity loss, primarily due to land-use changes, greenhouse gas emissions, and pollution throughout their value chains (Scope 3). The report reveals, for example, that over two-thirds of the total estimated biodiversity impact within the MSCI All Country World Index (ACWI)[5] is attributable to just 250 companies, with the top 10 alone contributing more than 15%. This concentration emphasises the need to focus engagement in the financial sector on key players. The report also shows that pollution (31%), land use (23%), and water use (12%) are the primary determinants of firms' impacts on biodiversity, and these factors vary significantly across sectors. The study highlights that sectoral impact rankings shift depending on whether absolute impacts or impact intensities (per revenue unit) are considered. Biodiversity loss is predominantly driven by value chain activities rather than direct operations, making a full-lifecycle, multi-driver approach essential for practical financial risk assessment and mitigation (Finance for Biodiversity Foundation, 2024).

Moving from impact to dependence, companies and sectors depend significantly on biodiversity, introducing a distinct but interconnected dimension of nature-related financial risk. The economic system, across all sectors, exhibits pervasive and differentiated dependencies on ecosystem services, which underpin core business functions and supply chain operations. According to the Finance for Biodiversity Foundation (2024),

[5] The MSCI ACWI is a global stock market index that gives investors exposure to thousands of companies across both developed and emerging markets, offering a simple way to track the performance of the world's equity markets in a single benchmark.

all industries depend to varying degrees on critical ecosystem services such as surface and groundwater, flood and storm protection, erosion control, and water flow maintenance. These dependencies are particularly pronounced in the food products and beverages sectors, where reliance on provisioning and regulating services is structurally embedded in upstream value chains (Scope 3). While these dependencies do not necessarily imply financial risk, the report highlights that the decline or shortage of ecosystem services can significantly disrupt production processes, increase costs, and weaken companies' ability to recover, especially in regions that are already stressed. Therefore, a company's geographic exposure and the locations of its suppliers are crucial in evaluating nature-related risks.

Regarding banks and financial institutions, the most notable impact on biodiversity is indirect, stemming from lending to companies and projects that harm biodiversity, as well as investing in assets and firms with negative environmental impacts. The consequences of biodiversity loss for borrowers translate into more risks and worse performance for banks. This is why, among the supervisory priorities for the coming years, 2024–2026, is the assessment and management of short-, medium-, and long-term natural and climate-related risks. Analysing supply chain risks and concentration risks, such as understanding customers' geographic presence, dependencies, and vulnerabilities, as well as scenario planning and the stress testing framework, are some of the activities banks are called upon to undertake in light of the increased frequency and severity of natural and climate-related events.

2.3 The Case of the Eurozone: A Call for Regulatory Attention

A recent Occasional Paper by the European Central Bank (Boldrini et al., 2023) provides one of the most detailed insights into the extent to which non-financial corporations (NFCs) and euro area banks rely on ecosystem services. Drawing on data from more than 4.2 million firms and over €4.3 trillion in corporate loans issued by 2500 banking groups, the study focuses on structural dependency, which is often invisible in traditional financial analysis.

The findings are striking: around 72% of these firms depend on at least one ecosystem service, such as pollination, water availability, or soil stability. However, this figure rises even further once indirect dependencies are taken into account, especially those transmitted through global supply chains. By linking firm-level data to the environmentally extended input–output database EXIOBASE,[6] the authors show that ecological risks are not limited to obvious sectors or locations. Even firms operating in low-impact regions may rely on suppliers in areas under severe environmental stress.

Crucially, these dependencies are not theoretical. They translate into real production risks. For example, declining crop yields resulting from the loss of pollinators, shortages of clean water due to aquifer depletion, or disruptions in raw materials caused by deforestation and soil erosion could all harm firms' operations.

This study is important for financial regulation because it connects ecological weakness to financial stability. The authors estimate that nearly 75% of the euro area's corporate loan portfolios, amounting to €3.24 trillion, are tied to firms with high dependency on nature. This means that environmental degradation is not just a matter of reputational risk or long-term sustainability; it is a clear and present concern for asset quality and financial stability.

These risks are geographically widespread but sectorally concentrated. Sectoral exposure is especially pronounced in the agriculture, construction, manufacturing, energy, and transportation industries. Moreover, the interconnected nature of supply chains means that localised ecological shocks can have far-reaching financial consequences.

In short, Boldrini et al. (2023) present a compelling case that the loss of biodiversity and the disruption of ecosystem services pose significant environmental, economic, and systemic financial risks, which warrant regulatory attention. The study invites a rethinking of how we define and measure risk in banking, pushing nature dependency closer to the heart of financial supervision in the euro area.

To estimate how the aforementioned ecological dependencies could materialise into credit risk, Boldrini et al. (2023) also conduct a sensitivity analysis using biodiversity loss scenarios and link them to potential increases in the probability of default (PD) for each firm. The results show that if the world follows a high-emission, low-mitigation pathway, banks'

[6] EXIOBASE is a detailed, environmentally extended input–output database that integrates economic and environmental information across more than 160 sectors and nearly 50 countries. It provides high-resolution data on how goods and services are produced, consumed, and traded globally, while quantifying the associated environmental pressures, including land use, emissions, and resource extraction (Stadler et al., 2018).

credit losses could be up to 2.73 times higher than in a sustainability-aligned scenario. In some countries, such as Germany, Ireland, Belgium, and Lithuania, losses could be as much as five times higher due to strong sectoral dependencies and localised biodiversity deterioration. The analysis also reveals that indirect dependencies through global supply chains are just as important as direct ones, suggesting that even well-capitalised or geographically diversified banks cannot assume insulation from ecological risk. The ECB finds a small inverse relationship between banks' capital adequacy, specifically their Tier 1 capital ratios, and ecological exposure. This suggests that banks with less capital might face greater risks from nature-related credit issues. Such exposure could increase systemic weaknesses during times of ecological or financial stress.

In sum, the ECB's report presents compelling evidence that biodiversity loss is not merely an environmental concern, but a latent financial and macroeconomic risk already embedded in credit market structure. The widespread and often hidden dependencies between economic actors and ecosystem services create a channel through which environmental degradation can rapidly undermine corporate solvency and financial stability. This reinforces the urgency for central banks, supervisors, and financial institutions to integrate nature-related risks into their risk assessment frameworks, portfolio allocation, and macroprudential policies.

2.4 The Economic Rationale Underlying the Financial Regulation

Financial regulators have turned increasing attention to the risks posed by biodiversity loss in recent years. This is not simply a matter of environmental concern; it is a response grounded in decades of economic thinking that demonstrates why markets alone are insufficient to protect ecosystems. Several core economic theories help explain why regulatory intervention is legitimate and necessary.

A primary reason lies in the concept of externalities. As early as the 1920s, Pigou observed that companies frequently impose costs on society, such as pollution and habitat destruction, without incurring those costs themselves (Pigou, 1920). Biodiversity loss fits this pattern perfectly. According to Pigou (1920), when a company creates costs for society but does not sustain them financially, the market does not function well because prices do not reflect the actual environmental impact. Since all share the benefits of biodiversity, but the costs of its destruction are not borne by those who damage it, this imbalance is addressed through regulatory measures such as taxes, compensation plans, and environmental standards.

A second reason regards information asymmetry. Reliable data is necessary for markets to operate effectively, but in the case of biodiversity, the data is frequently lacking, inconsistent, or incomplete. Investors may not be aware of a company's reliance on natural systems or the extent of harm it causes to these systems. This type of asymmetry can lead to mispricing of risk and misallocation of capital (Akerlof, 1970; Stiglitz, 2000). This may explain why regulations now require financial actors to disclose the impact of their portfolios or their exposure to biodiversity loss.

A third reason is linked to biodiversity as a global public good; yet, at the same time, no one, unless encouraged, has a sufficient incentive to invest in its conservation (free riding). Binding regulations, public funding, and reporting and transparency requirements are essential levers to counter free riding.

The relationship between managers and shareholders is another reason for intervention. Agency theory suggests that managers may prioritise short-term profits over long-term stability, particularly if their compensation depends on quarterly returns (Jensen & Meckling, 1976). Adding sustainability goals to governance frameworks or executive pay can alter incentives towards strategies that consider the impact on the natural environment. This way, regulation links private decision-making to public interest.

Finally, and perhaps most urgently, biodiversity loss is now understood to pose risks to financial stability itself. Studies by central banks and financial institutions have shown that the degradation of nature can threaten supply chains, create stranded assets, and amplify sectoral shocks, similar to a financial crisis or a climate catastrophe (DNB & PBL, 2020; NGFS, 2022). In this light, supervisors and central banks begin to integrate nature-related risks into their mandates, including through scenario analysis, prudential guidelines, and stress testing.

These insights from economics make a compelling case: left to their own devices, markets do not protect biodiversity, and in some cases, they actively undermine it. That is why regulation is not just a policy choice but an economic necessity.

2.5 From Global Biodiversity Agreements to Financial Regulation

The growing awareness of biodiversity loss as a systemic financial risk has led to a progressive translation of global biodiversity goals into regulatory and supervisory initiatives targeting the financial sector. While the Convention on Biological Diversity (CBD) entered into force on 29 December 1993, laying the foundational principles of biodiversity conservation, it is the Kunming-Montreal Global Biodiversity Framework (GBF), approved during the COP 15 held between 2021 and 2022, that most clearly articulates the new expectations for financial institutions. The GBF establishes an ambitious path to halt biodiversity loss and mobilise funding for the cause. It includes four long-term macro-objectives for 2050 (Protect and Restore, Prosper with Nature, Share Benefits Fairly, and Invest and Collaborate) and 23 targets to be achieved by 2030.

Among its 23 targets for 2030, Target 15 requires large companies and financial institutions to assess, disclose, and reduce their dependencies and impacts on biodiversity. Target 19 aims to mobilise at least USD 200 billion annually in biodiversity-related funding, underscoring the central role of private capital and financial actors in closing the biodiversity finance gap. These two concepts are explored in depth in the policy paper "*Aligning financial flows with the Kunming-Montreal Global Biodiversity Framework*", produced by *the United Nations Environment Programme Finance Initiative* (*UNEP FI*), which urges financial actors to integrate the *GBF* into decision-making processes in order to align investments with the 2030 targets and the 2050 objectives.

This shift is reflected in several regulatory and supervisory developments, particularly in the European Union, where biodiversity is being incorporated into the sustainable finance architecture.

2.5.1 Biodiversity in the EU Sustainable Finance Framework

In the European Union, biodiversity has become a key pillar of the sustainable finance agenda. Recognising that the preservation of ecosystems is an environmental imperative and a condition for economic and financial resilience, EU institutions have taken steps to embed biodiversity objectives into financial regulation and disclosure standards. A central element is the EU Biodiversity Strategy for 2030, approved in 2020. This strategy outlines a bold plan to protect at least 30% of the

Union's land and sea areas and to restore damaged ecosystems. The European Commission has pledged to allocate €20 billion annually to achieve these objectives, utilising EU funds, national resources, and private investment. The Strategy clearly states that protecting biodiversity and restoring nature cannot depend only on public or charitable funding. It emphasises that the involvement of financial intermediaries, such as banks, institutional investors, and asset managers, is crucial to directing money towards nature-friendly investments. As part of this strategic plan, the European Council also approved the first EU law on nature restoration, *the Nature Restoration Law*, which provides long-term recovery objectives for terrestrial and marine areas. The regulation aims to restore a resilient and biodiverse nature and to achieve the EU's climate mitigation and adaptation objectives. Here too, the regulation stresses the role of finance, highlighting how "every euro invested in nature restoration adds between 4 and 38 euros in benefits".[7] After months of deadlock, the EU Nature Restoration Law was politically agreed upon in June 2024. Its formal adoption was expected in the second half of 2024, with implementation beginning in 2025; however, Member States will have flexibility in defining their national restoration plans.

The EU adopted the Taxonomy Regulation to guide financial markets in identifying environmentally sustainable activities (Regulation EU 2020/852). The Taxonomy defines a standard economic activity classification system that substantially contributes to six environmental objectives, including the protection and restoration of *biodiversity and ecosystems*. An activity is considered "taxonomy-aligned" if it:

- provides a substantial contribution to at least one environmental objective;
- does no significant harm (DNSH) to any of the others;
- complies with minimum social safeguards;
- and meets the technical screening criteria established by the European Commission.

The Taxonomy Regulation and the related Delegated Acts, which specify technical screening criteria, enable financial institutions to identify and

[7] https://environment.ec.europa.eu/topics/nature-and-biodiversity/nature-restoration-law_en.

channel investments towards biodiversity-friendly activities, and integrate biodiversity performance into capital allocation, due diligence, and product labelling. These developments feed directly into corporate and financial reporting obligations. Companies are required by Article 8 of the Taxonomy to report the percentage of turnover, CAPEX, and OPEX that goes towards ecologically friendly initiatives, such as those that support biodiversity.

The scope and ambition of the EU Taxonomy-related disclosure requirements have been progressively expanded through a phased implementation process. The same degree of disclosure will be required of financial institutions in 2024, such as banks, insurers, and asset managers. Key performance indicators (KPIs) that demonstrate how well their investment and financing portfolios conform to the EU Taxonomy must be made public. Sharing information about how they are allocating funds to assets that adhere to taxonomy standards is part of this. They reaffirm their commitment to financing the green transition by taking this step. These revelations increase openness.

The disclosures provide investors, customers, and the general public with a better understanding of the environmental credibility of corporate strategies by demonstrating how closely companies align with EU sustainability goals. They also influence the choices made about investments. Investors can determine which companies genuinely contribute to sustainability goals and allocate their capital effectively with the aid of more comparable and transparent data. This could result in a reallocation of funds from detrimental endeavours to those that promote biodiversity and climate goals. These disclosure requirements also encourage accountability. They put pressure on businesses to improve their environmental performance by making them publicly report on their alignment—or lack thereof. Corporate change is ultimately driven by this regulatory transparency, which fosters a broader movement towards more environmentally friendly business practices.

The Corporate Sustainability Reporting Directive (CSRD) strengthens this requirement by mandating large and publicly listed companies to disclose risks and impacts related to the natural environment. The accompanying European Sustainability Reporting Standards (ESRS), specifically ESRS E4–Biodiversity and Ecosystems, require firms to outline their strategies for aligning with the GBF and the EU Biodiversity Strategy. The standard calls for disclosures on:

- dependencies on ecosystem services,
- material biodiversity-related risks,
- and transition plans to "nature-positive" business models.

The CSRD entered into force on 5 January 2023 and will apply in stages: from 1 January 2024 for companies already subject to the NFRD (reporting in 2025); from 1 January 2025 for large companies not previously subject; and from 1 January 2026 for listed SMEs. The implementation of the final ESRS E4 is postponed to 2026. Article 29 of the Loi Énergie-Climat in France requires financial institutions to disclose biodiversity-related risks starting from 2022, making France a forerunner in aligning finance with biodiversity concerns.

The Sustainable Finance Disclosure Regulation (SFDR) connects corporate disclosure to financial intermediaries. It requires fund managers and institutional investors to report how their portfolios match the EU Taxonomy, including details on biodiversity. The SFDR application commenced on 10 March 2021, while the Regulatory Technical Standards (RTS), which include Principal Adverse Impact (PAI) indicators such as "impact on biodiversity-sensitive areas", became applicable as of 1 January 2023. This way, the Taxonomy, CSRD/ESRS, and SFDR create a coherent data pipeline and accountability from companies to financial markets.

In February 2025, the European Commission adopted the "Omnibus Package", a broad simplification initiative aimed at reducing regulatory burdens and enhancing EU competitiveness.[8] The package introduces significant changes to the EU sustainable finance framework, including amendments to the Corporate Sustainability Reporting Directive (CSRD) and the EU Taxonomy. The overarching goal is to streamline sustainability reporting, lower compliance costs, and focus regulatory efforts on the largest companies with the greatest environmental and climate impact. The most substantial reform concerns the CSRD: around 80% of companies would be removed from its scope, with reporting obligations retained only for firms employing more than 1000 people and having either over €50 million in net turnover or a balance sheet total above €25 million. Additional measures include a voluntary sustainability reporting standard for companies in value chains with up to 1000 employees, a simplified

[8] https://finance.ec.europa.eu/news/omnibus-package-2025-04-01_en.

and revised set of ESRS, the elimination of sector-specific ESRS, and the maintenance of limited assurance requirements. Reporting obligations under the EU Taxonomy would likewise be restricted to the largest companies (at least 1000 employees and €450 million in turnover), with others allowed to report voluntarily. A separate "stop-the-clock" proposal postpones by 2 years the reporting deadlines for companies currently scheduled to report in 2026–27.

2.5.2 *Financial Supervision and Prudential Regulation*

Financial supervision has also evolved to address biodiversity risks. The European Banking Authority (EBA) and the European Central Bank (ECB) are progressively integrating biodiversity-related risks into their guidelines for financial institutions.

Over the past few years, European financial supervisors have gradually included climate and environmental factors, such as biodiversity, in their financial oversight. In 2020, the European Central Bank (ECB) released a supervisory guide detailing how banks should handle climate and environmental risks responsibly (ECB, 2020). Crucially, the guide recognised that threats such as ecosystem degradation and biodiversity loss pose real financial risks to banks and their clients. By March 2023, banks were expected to have completed materiality assessments covering climate and environmental risks. However, 28 banks received binding directives and potential sanctions from the ECB for failing to meet the deadline. The ECB's ambitious 2024–2025 Climate and Nature Plan[9] formally elevates nature risk to a third strategic pillar, highlighting the deep link between financial stability and healthy ecosystems.

Meanwhile, the European Banking Authority (EBA) has taken decisive steps to integrate environmental, social, and governance (ESG) risks into the prudential framework, recently releasing its first detailed ESG risk management guidelines in early 2025 (EBA, 2025). These guidelines go further than climate issues, explicitly requiring banks to identify and incorporate risks from ecosystem degradation and biodiversity loss into

[9] https://www.ecb.europa.eu/ecb/climate/our-climate-and-nature-plan/shared/pdf/ecb.climate_nature_plan_2024-2025.en.pdf.

key risk areas, such as sectoral exposure, scenario analysis, transition planning, and credit policies, while aligning with both the EU Taxonomy (e.g., via the Green Asset Ratio) and SFDR disclosure criteria, including key biodiversity impact indicators (PAI). Banks must embed these considerations in their corporate risk appetite frameworks, capital planning, and public disclosures. They must also set measurable targets related to nature, such as avoiding deforestation or restoring ecosystems. The latest EBA Guidelines (EBA, 2025) explicitly call for banks to assess how ecosystem degradation and biodiversity loss may affect traditional banking risks, such as credit, market, and operational risks. In particular, according to the EBA Guidelines, financial institutions should assess both transition and physical risk drivers, considering at least the following: for transition risks: a) "the main economic sectors that the financed assets support or in which the institution's counterparty has its principal activities"; b) "ongoing and potential future material changes in public policies, technologies and market preferences". For physical risks: a) "the geographical areas in which key assets of counterparties (e.g., production sites) and, in particular for real estate exposures, physical collateral is located"; b) "the vulnerability level to environmental hazards (e.g., temperature-related, wind-related, water-related, solid mass-related hazards) associated with different climate scenarios and transition pathways".

On the capital markets side, the European Securities and Markets Authority (ESMA) has focused on sustainable investment disclosures and combating greenwashing. Within its 2022–2024 roadmap, ESMA flagged biodiversity as an emerging risk beyond climate change. In mid-2023, its "Trends, Risks, and Vulnerabilities" report highlighted biodiversity as the "new frontier" in ESG investing, warning that its loss carries physical and transition risks, reinforcing climate shocks. Although biodiversity-focused investment funds (with around €850 million in assets by mid-2023) are still a niche market, two-thirds are classified under SFDR Article 9, signalling fast growth and the need for investor protection. Accordingly, ESMA released guidance in 2025[10] to ensure that sustainability claims, including those related to nature, are accurate, precise, and not misleading.

[10] https://www.esma.europa.eu/sites/default/files/2025-07/ESMA36-429234738_-154_Thematic_notes_on_clear__fair___not_misleading_sustainability-related_claims.pdf.

Although the European Insurance and Occupational Pensions Authority (EIOPA) regulates insurers, its pioneering work on nature-related risk has resonated through the EU's broader financial supervisory community. In March 2023, EIOPA released a staff paper[11] highlighting how biodiversity loss—such as deforestation or diminished pollination—can drive underwriting losses and investment risk. It also emphasised how insurers can actively support conservation by investing in sustainable projects or excluding harmful practices. Following this, EIOPA's June 2025 supervisory report became the first EU financial authority to map biodiversity risk practices in the insurance sector. It was found that while around 20% of insurers mention biodiversity in their ORSAs, most do so in a qualitative manner. Key challenges include a lack of shared global metrics, the local nature of ecological risks, and difficulty disentangling biodiversity from climate issues—yet the report lauded early adopters and urged deeper integration.

These initiatives demonstrate a clear shift: biodiversity and nature risk are no longer environmental side issues but core components of EU financial supervision. Across banks, insurers, and markets, Europe's regulators are increasingly expecting financial institutions to measure, manage, disclose, and ultimately respond, with real capital and policies, to the risks arising from biodiversity loss.

2.6 Conclusions

Ultimately, biodiversity loss is not only affecting nature; it is also starting to reshape how financial risks unfold. Banks and investors are already feeling the consequences, from the erosion of ecosystems to the knock-on effects of new regulations or public pressure. Still, the financial sector is not just on the receiving end. As this chapter has shown, it also holds real power to influence the outcome. When financial institutions adjust how they assess risks, set priorities, and steer investments, they can help shift the system towards one that protects and restores nature. Regulation and oversight are helping push this change, especially in Europe. However, the bigger challenge is to make nature a regular part of financial thinking,

[11] https://www.eiopa.europa.eu/publications/eiopa-staff-paper-nature-related-risks-and-impacts-insurance_en#:~:text=Protecting%20nature%E2%80%99s%20biodiversity%20and%20ecosystems,overall%20financial%20stability%20at%20risk.

not a box to tick, but something that genuinely shapes how capital is moved, how risk is priced, and how long-term value is defined.

References

Akerlof, G. A. (1970). The market for “lemons”: Quality uncertainty and the market mechanism. *Quarterly Journal of Economics, 84*(3), 488–500. https://doi.org/10.2307/1879431

Boldrini, S., Ceglar, A., Lelli, C., Parisi, L., & Heemskerk, I. (2023). *Living in a world of disappearing nature: Physical risk and the implications for financial stability*. ECB Occasional Paper (2023/333).

De Nederlandsche Bank (DNB) & PBL Netherlands Environmental Assessment Agency. (2020). *Indebted to nature: Exploring biodiversity risks for the Dutch financial sector*. Amsterdam & The Hague.

ECB. (2020). *Guide on climate-related and environmental risks: Supervisory expectations relating to risk management and disclosure*. European Central Bank, November 2020.

European Banking Authority. (2025). *Guidelines on the management of environmental, social and governance (ESG) risks*.

Finance for Biodiversity Foundation. (2024). *Assessment of the biodiversity impacts and dependencies of globally listed companies: A collaborative multi-tool footprinting approach*. https://www.financeforbiodiversity.org/publications/

Jensen, M. C., & Meckling, W. H. (1976). Theory of the firm: Managerial behaviour, agency costs and ownership structure. *Journal of Financial Economics, 3*(4), 305–360. https://doi.org/10.1016/0304-405X(76)90026-X

Johnson, J. A., Baldos, U., Hertel, T., Nootenboom, C., Polasky, S., & Roxburgh, T. (2020). *Global futures: Modeling the global economic impacts of environmental change to support policy-making*.

Network for Greening the Financial System (NGFS). (2022). *Central banking and supervision in the biosphere: An agenda for action on biodiversity loss, financial risk and system stability*. NGFS Occasional Paper.

Pigou, A. C. (1920). *The economics of welfare*. Macmillan.

Rimo, G., Cenni, S., & Cosma, S. (2024). Perdita di biodiversità, sistema finanziario e banche: regolamentazione e impatti attesi. *BANCARIA, 9*, 74–85. ISSN 0005-4623.

Stadler, K., Wood, R., Bulavskaya, T., Södersten, C. J., Simas, M., Schmidt, S., & Tukker, A. (2018). EXIOBASE 3: Developing a time series of detailed environmentally extended multi-regional input–output tables. *Journal of Industrial Ecology, 22*(3), 502–515. https://doi.org/10.1111/jiec.12715

Stiglitz, J. E. (2000). The contributions of the economics of information to twentieth century economics. *Quarterly Journal of Economics, 115*(4), 1441–1478. https://doi.org/10.1162/003355300555015

World Economic Forum. (2025). *The global risks report 2025* (20th ed.). World Economic Forum. https://www.weforum.org/publications/global-risks-report-2025/

CHAPTER 3

Banking Activity in the Age of Biodiversity Risk: Insights from Academic Literature

Abstract This chapter reviews the academic literature analysing the link between biodiversity loss and banking activity. By employing a bibliometric approach in conjunction with a systematic review, this chapter examines a research field that has been largely overlooked for over a decade but has experienced significant growth following key global events, such as the Kunming Declaration. The performance analysis reveals a rapidly expanding yet still fragmented literature, concentrated mainly in interdisciplinary sustainability journals rather than mainstream finance outlets. Science mapping identifies three thematic clusters: (i) interconnected ecological pressures and macro-financial vulnerabilities; (ii) bank-level exposure to biodiversity loss; and (iii) biodiversity disclosure and pricing of nature-related risks. Across clusters, the literature shows that biodiversity loss constitutes a financially material source of credit, market, operational, and systemic risk. However, significant gaps persist, including the lack of standardised disclosure metrics, limited quantification of transmission channels, and underdeveloped prudential approaches. The chapter concludes by outlining a future research agenda aimed at supporting the incorporation of nature-related risks into banking practice and financial supervision.

Keywords Biodiversity finance · Banking · Financial institutions · Bibliometric analysis · Systematic review

G. Rimo and S. Cosma, *Biodiversity and Banks*, Palgrave Studies in Impact Finance, https://doi.org/10.1007/978-3-032-17631-8_3

3.1 Introduction

In the previous chapter, we have shown how biodiversity loss is increasingly recognised as a material driver of economic and financial risk. We also clarified that, in advanced economies such as the Eurozone, banks and financial intermediaries are exposed to these risks not only through the impact that ecosystem degradation exerts on the real sector, but also through regulatory constraints, technological transitions, and shifting market expectations. Against this background, a central question emerges: *How has academic research examined the relationship between banking and biodiversity loss, and what conclusions has it reached? Which gaps remain open?*

Previously, numerous academics have noted a general disinterest in the academic community concerning the financial impacts of biodiversity loss, urging researchers in the fields of banking and finance to take action (Cosma et al., 2023; Karolyi & Tobin-de la Puente, 2023). As suggested by Giglio et al. (2025), biodiversity risk has mainly been neglected by researchers in economics and finance due to its complexity and the associated difficulties in measuring and quantifying its various elements.

The numerous calls for action launched recently have caused quite a stir in the academic community, leading to the publication of several articles analysing the issue of biodiversity loss from a financial perspective. However, these articles remain fragmented and scattered, failing to provide an overview of the phenomenon.

This chapter aims to systematise these disaggregated contributions, thus providing a careful examination of the primary scientific literature via bibliometric techniques and a structured content review. Thus, the aim is to provide a structured map of the academic contributions that have progressively explored the nexus between ecosystems, financial risk, and banking stability.

3.2 Data and Methodology

3.2.1 Sample Selection

The bibliographic dataset was constructed using Scopus, which is widely recognised as one of the most comprehensive databases for academic research in the social sciences (Pattnaik et al., 2020). We looked for all studies that had the terms "*biodiversity*" AND ("*bank**" OR "*financial* *institution**") in the title, abstract, or keywords in order to find the pertinent literature at the nexus of biodiversity and banking (Bahoo, 2020; Cosma et al., 2023). In line with previous bibliometric studies in sustainable finance (e.g., Pandey et al., 2024), we restricted the search to the subject areas "Economics, Econometrics, and Finance" and "Business, Management, and Accounting". To ensure comparability and academic rigour, we refined our sample by including only articles, reviews, books, and book chapters and limiting results to publications written in English (Cosma et al., 2023).

An initial dataset of 168 records resulted from this process. Then, each document underwent a manual screening process to determine how well it aligned with the review's specific goals. This process involved reading abstracts, keywords, and, if needed, complete texts. After cleaning and eligibility assessment, the final sample consists of 36 documents published in 30 different sources between 2009 and 2025. Together, these papers constitute the main body of scholarly work that connects banking and biodiversity loss.

3.2.2 *Methodology*

The selected studies were examined through a bibliometric methodology, complemented by a systematic content review. The bibliometric approach is particularly suitable for mapping emerging research domains, as it involves the application of quantitative and statistical techniques to bibliographic data, including publication metadata and citation patterns (Mukherjee et al., 2022). This dual perspective enables both a descriptive overview of the field and an in-depth exploration of its intellectual and conceptual structure.

Specifically, the bibliometric analysis included performance analysis and science mapping (Cosma et al., 2023). Performance analysis relies on activity indicators such as the number of publications and citations to assess productivity and impact within a research domain. In this study, we examined trends in publication and citation counts, as publication volume is commonly considered a proxy for scientific productivity, while citation frequency reflects scholarly influence and visibility (Donthu et al., 2021). Through this lens, we identified the most prominent authors, sources, and countries contributing to the literature on biodiversity and financial institutions.

In contrast, science mapping seeks to identify knowledge clusters and offer a spatial representation of the relationships between scientific components (McCain, 1990). We employed keyword analysis to map the relationships between publications, thereby revealing the conceptual structure of the field. We specifically looked at the co-occurrence of authors' keywords (Callon et al., 1983). This method assumes that when two terms frequently appear together in the same documents, the underlying concepts they represent are cognitively linked (Donthu et al., 2021). The resulting maps enabled us to identify core research themes, emerging topics, and the network of conceptual interconnections shaping the field.

All bibliometric analyses were carried out using VOSviewer for network visualisation and Bibliometrix/Biblioshiny (R package) for data processing (Aria & Cuccurullo, 2017; van Eck & Waltman, 2010).

3.3 Performance Analysis

3.3.1 Publication Pattern

Figure 3.1 enables us to understand how the scientific literature on biodiversity and banking, as well as its citation impact, has evolved. The temporal dynamics displayed in the figure highlight a research field that remained marginal and sporadically explored for more than a decade, followed by a sudden and marked surge of interest after 2021.

The first pioneering articles on the subject are those by Vávrová (2009) and Randalls (2009). The article by Vávrová (2009) analyses the EU's *Environmental Liability Directive* (ELD, 2004/35/EC), illustrating how biodiversity loss can translate into a new class of operational and liability risks for financial institutions. According to the author, the ELD introduces a category of risk (*damage to biodiversity*) that is difficult to insure, forcing financial institutions (especially insurance and reinsurance companies, but also banks that provide guarantees) to develop new risk management models, new financial products, and new risk assessment processes, in line with Solvency II. In this sense, biodiversity risks enter the balance sheets of financial institutions through both liability channels and prudential regulation, compelling insurers, reinsurers, and banks to reassess risk modelling practices, capital planning, and product design. The ELD thus represents an early regulatory mechanism through which nature-related risks become financially material, anticipating later

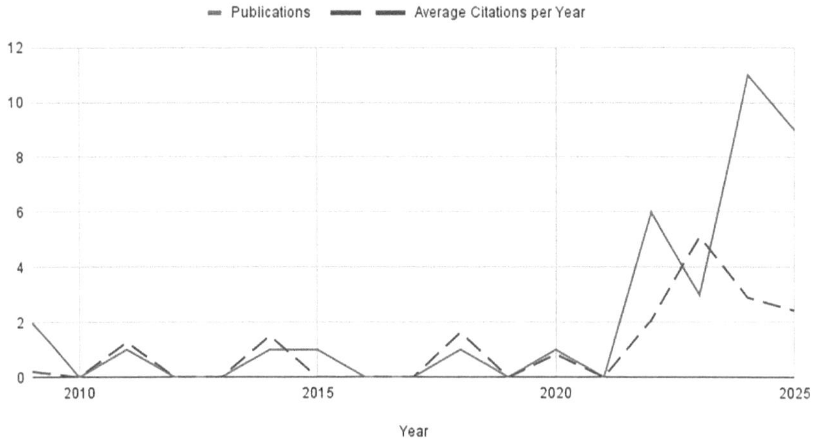

Fig. 3.1 Publications and average citations per year (Authors' elaboration)

debates on integrating environmental externalities into financial stability frameworks.

Randalls (2009) offers one of the first systematic analyses of the financialization of environmental risks, showing how banks, insurance companies, and other intermediaries transform climatic and environmental events, as well as elements of biodiversity, into quantifiable and negotiable risks. The author documents the growth of instruments such as weather derivatives, catastrophe bonds, carbon markets, and biodiversity offsets, highlighting how the environment is made measurable, modellable, and therefore incorporable into financial processes. Although the focus is on urban areas and climate risks, the chapter clearly anticipates the idea, now central to the debate on ESG and nature-related risks, that degraded ecosystems represent sources of financial risk and that financial institutions play a key role in managing them through innovative products and risk pricing mechanisms.

Between 2009 and 2019, there were only sporadic and isolated contributions, and the average annual citations and number of publications remained near zero. According to this pattern, the relationship between biodiversity and financial institutions was still developing in the 2010s and was largely overshadowed by more general environmental issues, such as climate risk.

Following 2019, we observe an acceleration that coincides with a series of global policy and regulatory milestones, which have placed biodiversity squarely on the financial agenda. In particular:

- the *Kunming Declaration* (2021), which formally recognised biodiversity loss as an urgent global challenge and called for mainstreaming biodiversity considerations across economic and financial systems;
- the rapid development of the TNFD (Taskforce on Nature-related Financial Disclosures) framework (from 2021 onward), which signalled to banks and investors that nature-related risks would soon require structured identification, assessment, and disclosure;

These shifts acted as catalysts, generating heightened academic and policy attention. Therefore, the 2023–2024 publication peak represents the point at which biodiversity finance evolved from a developing field of study to a recognised one, bolstered by higher regulatory standards and more precise methodological instruments (e.g., CBF, ENCORE, biodiversity footprint models).

The corresponding rise in citations indicates that the latest contributions quickly became reference points, suggesting the field had entered a take-off and consolidation phase.

Overall, the pattern highlights how external regulatory milestones, particularly those of Kunming and the TNFD, played a pivotal role in driving research at the intersection of biodiversity and financial systems.

3.3.2 Most Impactful Contributors

Table 3.1 highlights the most prolific and impactful sources on the topic, in terms of the number of publications and citations, respectively. Although the review focuses on the interconnection between biodiversity and banking, the first result that emerges is the almost total absence of finance journals in the rankings. Overall, the ranking shows that the intellectual structure of this field is currently shaped more by economics (*Ecological Economics*, *Applied Economics Letters*) or interdisciplinary sustainability journals (*Business Strategy and the Environment*, *Journal of Cleaner Production*) than by mainstream finance outlets. This confirms that biodiversity-related financial research, while rapidly growing, is still transitioning from a peripheral niche towards a recognised sub-area within sustainable finance and environmental economics.

The author-level results in Table 3.2 clearly reveal that this is a young and still fragmented field, as almost all scholars in the sample have contributed only a single publication. With the exception of *Nedopil*, who appears with two documents, no author has produced more than one study on the topic. This pattern suggests that research on biodiversity in banking has not yet established a stable or cohesive academic community; instead, contributions are typically irregular and often originate from broader research agendas in sustainability, environmental economics, or green finance.

In terms of impact, the most cited authors, Albu, Chang, Mirza, Su, and Umar, each have 43 citations, all of which come from their co-authored article, "Resource extraction, greenhouse emissions, and

Table 3.1 Most impactful sources

Source	*NP*	*Source*	*TC*
Ecological economics	4	Business strategy and the environment	42
Extinction governance, finance, and accounting	3	Ecological economics	30
Journal of cleaner production	2	Journal of cleaner production	28
Adaptive strategies for green economy and sustainability policies	1	Corporate social responsibility and environmental management	5
Agricultural economics	1	Economics and environment	3
Applied economics letters	1	Economics letters	3
Business strategy and the environment	1	Environmental economics	3
Corporate social responsibility and environmental management	1	Finance research letters	3
CSR, sustainability, ethics, and governance	1	Adaptive strategies for green economy and sustainability policies	1
Economics and environment	1	Agricultural economics	1
Economics letters	1	Applied economics letters	1
Environmental economics	1	Extinction governance, finance, and accounting	1
Finance research letters	1	International review of financial analysis	1
...	...	...	...

Note The table provides an overview of the most prolific (Columns 1–2) and most cited (Columns 3–4) sources on the topic. *Source* Authors' elaboration

Table 3.2 Most impactful authors

Author	*Citations*	*Documents*
Albu L.L.	43	1
Chang T.	43	1
Mirza N.	43	1
Su C.-W.	43	1
Umar M.	43	1
Nedopil C.	42	2
Adamolekun G.	19	1
Koellner T.	19	1
Mulder I.	19	1
Singh S.	18	1
Tara K.	18	1
...	...	...

Note The table provides an overview of the most cited (Column 2) and most prolific (Column 3) authors on the topic. *Source*: Authors' elaboration

banking performance", published in *Resources Policy*. The concentration of citations around a few high-impact papers emphasises how early anchor studies influenced the development of new fields of study. Although these authors have only written one paper on the subject, their work has become a crucial point of reference for subsequent studies.

Overall, the author ranking reinforces the view that this is an early-stage research area, where influence is currently driven by individual seminal papers rather than by established networks of recurring contributors. As the field matures, we may expect greater continuity, larger co-authorship structures, and the formation of specialised research groups focused on biodiversity-related financial risks.

3.4 Science Mapping

To uncover the conceptual structure of the literature and identify the main thematic clusters within the field, we conducted a keyword co-occurrence analysis based on the Authors' keywords. This technique is widely used in bibliometric research to map the interlinking of ideas and concepts within a scientific domain. The underlying assumption is that when two keywords frequently appear together in the same publication, the concepts they represent are likely to be cognitively or thematically connected (Donthu et al., 2021).

The analysis was performed using VOSviewer, which applies a text-mining algorithm to generate visual keyword maps. In these maps, each keyword is represented by a node (a circle), whose size reflects its frequency in the set of Authors' keywords across the sample. Keywords that appear more frequently are displayed as larger nodes, signalling their centrality within the research domain.

The distance between nodes serves as a proxy for the strength of their association: keywords that co-occur more often are positioned closer together, while those with weaker or infrequent co-occurrences are placed farther apart (Laudano et al., 2018). VOSviewer also uses distinct colours to group related keywords into clusters, allowing us to identify the dominant research themes, emerging topics, and the conceptual sub-fields that characterise the intersection between biodiversity and financial institutions.

This mapping approach provides a structured and visually intuitive representation of the intellectual landscape of the field, complementing performance analysis and enabling a deeper understanding of how research topics evolve.

The result of the keyword co-occurrence analysis is graphically presented in Fig. 3.2.

The keyword co-occurrence analysis reveals three well-defined thematic clusters that together outline the conceptual landscape of the literature.

- The **first cluster** (***red***) groups keywords related to large-scale ecological pressures, such as biodiversity loss, deforestation, land-use change and climate vulnerability. It captures studies examining how these systemic environmental dynamics interact with financial system structures, including financial inclusion and governance quality, to generate macro-financial risks and development challenges.
- The **second cluster** (***green***) captures research on how banks and financial institutions are financially exposed to biodiversity loss and ecosystem-service dependencies. Keywords such as *biodiversity risks*, *banking*, and *risk assessment* point to studies examining how lending to nature-intensive sectors and portfolio dependencies on ecosystem services translate into credit risk, bank performance, and financial stability vulnerabilities.
- Finally, the **third cluster** (***blue***) captures research on biodiversity disclosure, transparency, and the pricing of nature-related risks. Keywords highlight studies on how banks integrate biodiversity

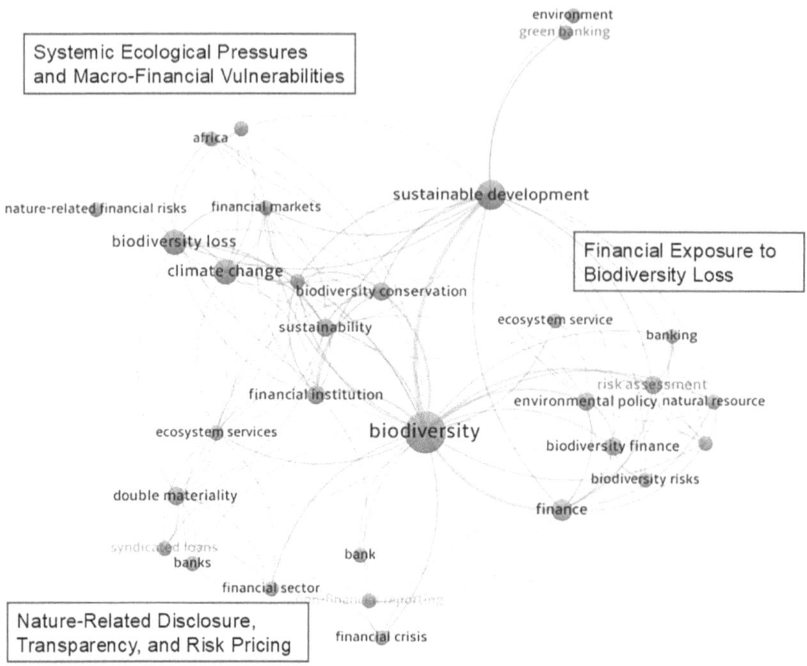

Fig. 3.2 Co-occurrence of keywords (Authors' elaboration)

information into loan pricing and reporting practices, as well as the limitations of current biodiversity metrics, disclosure frameworks, and measurement tools.

3.5 Discussion of Thematic Clusters

3.5.1 Red Cluster: Systemic Ecological Pressures and Macro-Financial Vulnerabilities

The Red Cluster adopts a broad systemic perspective, showing how biodiversity decline interacts with climate risks, deforestation, land-use change, and financial system characteristics, such as financial inclusion and governance quality, to generate macroeconomic and macro-financial vulnerabilities.

Several contributions emphasise the joint financial implications of biodiversity loss and climate vulnerability for firms and financial systems. Adamolekun (2024) shows that biodiversity-related exposures significantly increase the probability of corporate financial distress, with climate susceptibility exacerbating bankruptcy risks. These effects are amplified by firm-specific characteristics such as financial constraints, growth opportunities, and carbon-intensive business models, underscoring the interconnected nature of environmental degradation and corporate fragility. Cleary and Willcott (2022) quantify the physical costs of climate change in Canada, including biodiversity loss. Their findings indicate that early investments in emission reduction can substantially reduce long-term physical damages, reinforcing the economic rationale for proactive mitigation policies.

A second thematic strand focuses on the failures of environmental governance within large-scale development initiatives. Hardiman (2020) examines the Belt and Road Initiative,[1] arguing that banks and financing mechanisms have systematically underweighted environmental considerations in project evaluation. As a result, BRI corridors have intensified pollution, deforestation, biodiversity loss, and climate impacts. Linked to this macro-ecological perspective, Lunku et al. (2025) investigate how financial inclusion affects the dynamics of deforestation in Sub-Saharan Africa. Their results indicate that increased access to financial institutions can reduce environmentally harmful forest conversion, whereas unregulated financial markets may exacerbate degradation. This suggests that the structure of financial systems can either mitigate or accelerate environmental pressures, particularly in low-income regions where land-use change is a primary driver of biodiversity loss. In a similar vein, Sami et al. (2025) offer recommendations on the role of green finance and green banking in supporting sustainable development through initiatives such as green bonds, green loans, and low-carbon mutual funds.

Across all these works, a clear standard message emerges: biodiversity loss does not operate in isolation but interacts with climate change,

[1] The Belt and Road Initiative (BRI), launched by China in 2013, is a large-scale international development strategy aimed at expanding cross-border connectivity through investments in transport, energy, and digital infrastructure across Asia, Africa, and Europe. Although presented as a development initiative, the BRI has raised concerns regarding debt sustainability, environmental impacts, and the governance standards of funded projects.

deforestation, land-use pressures, and governance failures to generate multidimensional risks for firms, financial institutions, and economies.

3.5.2 *Green Cluster: Financial Exposure to Biodiversity Loss*

The Green Cluster brings together studies that analyse how biodiversity loss and portfolio dependencies on ecosystem services translate into financial risks for banks and the broader financial system.

A first group of papers highlights the direct financial consequences of lending to nature-intensive firms. Su et al. (2022) demonstrate that in emerging economies, banks that lend to high-emission, resource-dependent borrowers face a higher likelihood of default, suggesting that biodiversity-related vulnerabilities can compromise bank resilience. Complementing this perspective, Stolbov et al. (2025) document a bi-directional relationship between biodiversity conditions and the incidence of financial crises: ecological decline increases systemic fragility, while financial instability, in turn, exacerbates biodiversity loss.

The strong relationship between biodiversity and financial stability lies in the heavy dependence of financial institutions on ecosystem services, which exposes them to physical and transition risks. Mundaca and Heintze (2024) estimate that a significant share of European banks' equity portfolios is tied to firms with high ecosystem-service dependence, particularly in water-related services, indicating that banks are not only financiers of nature-dependent activities but are also vulnerable to ecosystem degradation themselves. Hadji-Lazaro et al. (2024) arrive at similar conclusions using the ENCORE framework and the Global Biodiversity Score. They indicate that a significant portion of the securities portfolios held by French financial institutions is exposed to both high dependencies on ecosystem services and substantial biodiversity footprints. The authors estimate that this exposure has an impact equivalent to the loss of approximately 130,000 km^2 of pristine natural areas.

These dependencies have long been underestimated. More than 10 years ago, Mulder and Koellner (2011) used a questionnaire submitted to 50 commercial banks to document that only five banks had taken significant measures to consider biodiversity risks and opportunities at the organisational level and within their loan portfolios. In particular, according to their survey, banks believed that their sector was exposed to biodiversity mainly through reputational risk. More recently, Stevenson (2024) has focused on bank financing of industrial animal farming, which has a very high impact on biodiversity, showing that credit often continues to flow towards activities associated with deforestation, biodiversity loss, and soil degradation.

Nedopil (2023) examines why biodiversity remains absent mainly from financial decision-making, despite mounting evidence of its economic and systemic relevance. Using the classic financial framework of costs, revenues, and risks, the paper identifies three structural barriers: the failure to internalise the costs of biodiversity loss due to incomplete property rights; the difficulty of monetising conservation benefits because of non-existent markets and pervasive free-rider problems; and the lack of standardised metrics to assess inherently local and non-linear biodiversity risks. From these gaps, the author derives four guiding principles: the need for regulatory limits on the exploitation of nature, a primarily local approach to biodiversity-related risk assessment, scaling biodiversity finance by leveraging climate-related co-benefits, and a more proactive role for financial institutions in shaping standards and instruments. Overall, the paper provides a conceptual foundation for understanding the biodiversity finance gap and outlines the conditions under which ecosystem protection can become "bankable". Van 't Klooster and Prodani (2025) also note that, while climate risks have become manageable for supervisory authorities, biodiversity risks remain difficult to standardise, leading to fragmented prudential approaches that are heavily dependent on national regulatory will.

The Green Cluster, therefore, highlights both the financial significance of biodiversity and the need for more robust tools and regulatory frameworks to incorporate nature-related risks into mainstream financial analysis.

3.5.3 *Blue Cluster: Nature-Related Disclosure, Transparency, and Risk Pricing*

The Blue Cluster groups together studies that investigate biodiversity-related risks through the lens of financial disclosure, transparency, and loan pricing, highlighting how banking institutions are increasingly integrating nature-related information into their credit decisions and reporting practices.

A first strand within this cluster concerns the pricing of biodiversity risk in syndicated loan markets, captured by Becker et al. (2025) and Li et al. (2025). Becker et al. (2025) demonstrate that lenders' exposure to biodiversity varies significantly across European countries, depending on borrowers' activities and the composition of their loan portfolios. Their evidence reveals that banks charge higher loan spreads when borrowers are more exposed to biodiversity loss, suggesting that creditors already internalise nature-related information as part of risk assessment. Li et al. (2025) strengthen this result by exploiting the Kunming Declaration as an exogenous regulatory shock. They find that, in the post-Kunming period, firms with high biodiversity risk face significantly higher loan spreads, thus documenting the emergence of a biodiversity risk premium. Moreover, their results suggest that this premium is more substantial for financially weaker borrowers and more pronounced among banks committed to ecological sustainability. In a similar vein, Barning et al. (2024) find that nature-related risks are financial risks priced in corporate bonds, highlighting the need for banks to incorporate these considerations into their risk management frameworks, encompassing identification, assessment, mitigation, and disclosure. Together, these studies illustrate how concerns about biodiversity propagate from the real economy to financial contracts, influencing credit conditions, maturities, loan amounts, and fees.

A second group of papers examines the disclosure and transparency practices of financial institutions regarding biodiversity. Azizi et al. (2025) analyse non-financial reports in the European financial sector and document substantial heterogeneity in both the scope and quality of biodiversity-related disclosures. While frameworks such as the EU Taxonomy, the SFDR, and the CSRD are pushing banks towards more structured reporting, the authors find that biodiversity remains considerably less disclosed than climate-related issues, reflecting an early stage of institutionalisation. Similarly, Broniewicz et al. (2024) assess the alignment of Polish banks with the new ESRS requirements, concluding that environmental disclosure practices remain fragmented, often inaccessible, and insufficient to meet the expectations for biodiversity reporting. They also highlight that banks tend to report extensively on climate, while biodiversity and ecosystem impacts are the least developed areas of disclosure.

Zhu et al. (2024) highlight a further weakness in the disclosure landscape. By evaluating the suitability of existing business biodiversity indices, tools intended to quantify firms' biodiversity impacts and dependencies for reporting and risk management purposes, they show that these metrics are poorly aligned with the requirements of the Kunming-Montreal Global Biodiversity Framework (GBF). Their gap analysis reveals that several GBF-critical dimensions, including ecosystem integrity, connectivity, restoration, aquatic and genetic diversity, nature-based solutions, and Indigenous knowledge, are not captured by current tools. As a result, firms and financial institutions cannot measure or disclose biodiversity impacts in a manner consistent with global expectations. The study, therefore, calls for integrated, updated indices that reflect the GBF's scope and support credible biodiversity reporting in line with both GBF and TNFD frameworks.

Thus, the Blue Cluster demonstrates both the progress made in recognising biodiversity as a financial variable and the structural weaknesses that still characterise nature-related reporting, ultimately pointing to an area where regulation, market incentives, and institutional capacities have not yet fully converged.

3.6 New Trends and Future Research Directions

In Fig. 3.3, we present an overlay analysis, which reveals the temporal stratification in the literature on biodiversity and banking (Raimondi et al., 2025). The darker nodes represent earlier research areas, while the lighter nodes indicate more recent conceptual developments. Overall, the visualisation suggests a progressive shift from broad environmental themes towards more specialised, finance-oriented notions of biodiversity risk.

The overlay visualisation shows that the field has rushed from broad sustainability themes towards more finance-oriented notions. Over time, the focus shifts towards more finance-specific concepts, where biodiversity begins to be linked to *financial institutions*, *ecosystem services*, *non-financial reporting*, and *syndicated loans*.

These newer themes demonstrate that biodiversity is now increasingly recognised as a financial risk factor, relevant to loan pricing, portfolio exposures, and disclosure frameworks (e.g., TNFD, ESRS).

However, even though these concepts are recently emerging in the literature, they remain weakly connected, indicating that the financial implications of biodiversity loss have only been partially explored. This suggests that future research should focus on integrating these emerging strands and addressing the structural gaps identified in the three clusters: incomplete disclosure metrics (Blue Cluster), limited understanding of bank-level exposures (Green Cluster), and macro-ecological mechanisms linking environmental degradation and financial stability (Red Cluster).

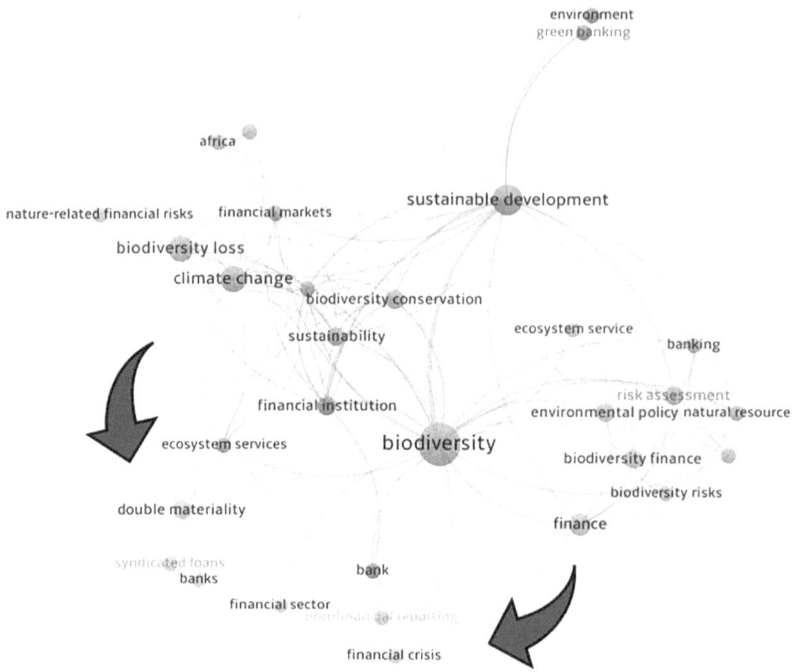

Fig. 3.3 Overlay analysis (Authors' elaboration)

Given the evidence of the negative impact on biodiversity of unregulated financial markets and, on the other hand, the effect of financial exclusion that can also increase pressure on ecosystems, future research should examine whether fintech-driven forms of financial inclusion (e.g., mobile banking, digital lending, blockchain-based microcredit) mitigate or exacerbate deforestation and biodiversity loss in regions where traditional financial access influences land-use decisions (Cosma & Rimo, 2023).

A promising avenue for future research concerns the development and evaluation of financial instruments designed to reduce the vulnerability of biodiversity-dependent economies (agriculture, fisheries, and forestry) to ecological degradation. These sectors are structurally exposed to nature-related shocks, yet the financial architecture supporting them remains underdeveloped. Scholars could investigate how emerging forms of *adaptation finance*, including agri-insurance schemes, biodiversity-linked loans, and nature-based solutions financing through blended-finance structures, can provide effective risk-transfer mechanisms or strengthen local adaptive capacity. A crucial question is under what conditions these instruments genuinely reduce ecological and financial fragility, rather than merely shifting risks across actors or generating moral hazard. Likewise, future studies could explore how to design incentives that reward sustainable land-use practices, discourage ecosystem-damaging activities, and channel capital towards community-level resilience initiatives. Understanding the interplay between financial innovation, ecological thresholds, and social inclusion will be crucial for ensuring that these tools contribute to both environmental protection and long-term economic stability in regions that depend on biodiversity.

The review's articles collectively demonstrate that banks face significant financial risks due to biodiversity loss and reliance on ecosystem services (e.g., Adamolekun, 2024). However, further research is needed to quantify the transmission channels that link bank-level risks to biodiversity degradation. Future studies could specifically model how disruptions to ecosystems affect credit, market, and operational risks for various banking business models.

Furthermore, one issue that has clearly emerged from this analysis is the underestimation of the dependence of financial systems on ecosystem services, which exposes banks to significant risks from biodiversity loss (Mundaca & Heintze, 2024). Technological innovation offers novel opportunities in this field. Solutions such as satellite monitoring, geospatial analytics, and blockchain-based traceability could enhance biodiversity-related due diligence and mitigate information asymmetry. Future research could investigate whether such technologies enhance loan pricing accuracy and mitigate portfolio exposure to natural resource-dependent sectors.

Another promising direction concerns the effectiveness and comparability of disclosure frameworks. Current reporting practices remain highly uneven across institutions, countries, and disclosure standards (Azizi et al., 2025). Research should assess whether the EU Taxonomy, CSRD, and ESRS are improving the consistency and decision-usefulness of biodiversity information, and how these frameworks intersect with global initiatives such as the TNFD and the Kunming–Montreal GBF. An important question is whether better disclosures actually lead to improved risk management, capital reallocation, or reductions in biodiversity-harmful financing.

Finally, studies like Zhu et al. (2024) highlight substantial gaps in biodiversity measurement tools. This opens avenues for developing and validating new biodiversity indices that align with GBF targets and TNFD metrics. Key open questions include how to integrate ecosystem integrity, connectivity, and Indigenous knowledge into quantitative models, as well as how financial institutions can operationalise these dimensions in portfolio analysis, stress testing, and scenario design.

3.7 Conclusions

This chapter has shown that scholarly research on the risks biodiversity poses to banking stability is still in its early stages but is evolving rapidly. The bibliometric data indicate that this area of research remained largely overlooked for over 10 years. It only began to gain traction following significant policy shifts, such as the Kunming Declaration and the establishment of the TNFD framework, which helped to secure a place for nature on the financial agenda. Despite this progress, the field remains underdeveloped. Disciplinary boundaries persist, with contributions from researchers scattered, and mainstream finance journals largely absent from the debate.

The literature's conceptual framework identifies three complementary yet unevenly advanced research domains: (i) systemic environmental pressures influencing macro-financial vulnerabilities; (ii) bank-level exposures associated with ecosystem-service dependencies and sectoral lending trends; and (iii) nascent studies on disclosure quality and the pricing of biodiversity risk. These clusters indicate that an increasing number of people are realising that losing biodiversity is not only an environmental disaster but also a source of credit, market, operational, and systemic risk for banks and other financial institutions.

The analysis also reveals that the current situation has several issues, including non-standardised disclosure metrics, limited quantification of bank-specific transmission channels, and insufficient integration of biodiversity risks into prudential frameworks. Measurement gaps persist as a significant impediment, hindering academic advancement and the capacity of financial institutions to manage nature-related risks effectively.

Ultimately, this chapter underscores the need for a more coherent, finance-focused research agenda that can effectively connect ecological science, risk modelling, and regulatory practice. To help supervisors, guide bank strategies, and ensure that the financial system supports global biodiversity goals instead of exacerbating ecological fragility, it will be crucial to strengthen the methodological foundations of biodiversity finance.

References

Adamolekun, G. (2024). Firm biodiversity risk, climate vulnerabilities, and bankruptcy risk. *Journal of International Financial Markets, Institutions and Money, 97*, Article 102075.

Aria, M., & Cuccurullo, C. (2017). Bibliometrix: An R-tool for comprehensive science mapping analysis. *Journal of Information, 11*(4), 959–975.

Azizi, L., Scope, C., Ladusch, A., & Sassen, R. (2025). Biodiversity disclosure in the European finance sector. *Ecological Economics, 228*, Article 108430.

Bahoo, S. (2020). Corruption in banks: A bibliometric review and agenda. *Finance Research Letters, 35*, Article 101499.

Barning, O., Broeders, D., De Jonge, M., Tiems, I., Verhoeven, N., & Van Wijmen, C. (2024). Nature-related financial risks and central bank risk management. *Journal of Risk Management in Financial Institutions, 17*(3), 316–331.

Becker, A., Di Girolamo, F. E., & Rho, C. (2025). Loan pricing and biodiversity exposure: Nature-related spillovers to the financial sector. *Research in International Business and Finance, 75*, Article 102724.

Broniewicz, E., Jastrzębska, E., & Lulewicz-Sas, A. (2024). Environmental disclosures according to ESRS in ESG reporting of selected banks in Poland. *Economics and Environment, 88*(1), 719–719.

Callon, M., Courtial, J.-P., Turner, W. A., & Bauin, S. (1983). From translations to problematic networks: An introduction to co-word analysis. *Information (International Social Science Council), 22*(2), 191–235.

Cleary, S., & Willcott, N. (2022). The cost of delaying to invest: A Canadian perspective. *Finance Research Letters, 50*, Article 103242.

Cosma, S., & Rimo, G. (2023). Fintech, financial inclusion, and social challenges: The role of financial technology in social inequality. In *Fintech and sustainability: how financial technologies can help address today's environmental and societal challenges* (pp. 107–128).

Cosma, S., Rimo, G., & Cosma, S. (2023). Conservation finance: What are we not doing? A review and research agenda. *Journal of Environmental Management, 336*, Article 117649.

Donthu, N., Kumar, S., Mukherjee, D., Pandey, N., & Lim, W. M. (2021). How to conduct a bibliometric analysis: An overview and guidelines. *Journal of Business Research, 133*, 285–296.

Giglio, S., Kuchler, T., Stroebel, J., & Zeng, X. (2025). Biodiversity risk. *Review of Finance, 2025*, rfaf063.

Hadji-Lazaro, P., Salin, M., Svartzman, R., Espagne, E., Gauthey, J., Berger, J., & Vallier, A. (2024). Biodiversity loss and financial stability as a new frontier for central banks: An exploration for France. *Ecological Economics, 223*, Article 108246.

Hardiman, R. (2020). Environmental considerations of the Belt and Road Initiative. In *The belt and road initiative: An old archetype of a new development model* (pp. 173–201). Springer.

Karolyi, G. A., & Tobin-de la Puente, J. (2023). Biodiversity finance: A call for research into financing nature. *Financial Management, 52*(2), 231–251.

Laudano, M. C., Marzi, G., & Caputo, A. (2018). A decade of the international journal of entrepreneurship and small business: A bibliometric analysis. *International Journal of Entrepreneurship and Small Business, 33*(2), 289–314.

Li, J., Jin, Y., & Zhou, P. (2025). Do banks price firms' biodiversity risk? Evidence from the Kunming declaration. *International Review of Financial Analysis, 2025*, 104557.

Lunku, H. S., Li, Z., & Danga, M. (2025). Deforestation and financial inclusion in sub-Saharan Africa: Evidence from the conditional and unconditional quantile regression approach. *Journal of Forest Economics, 40*(4), 297–326.

McCain, K. W. (1990). Mapping authors in intellectual space: A technical overview. *Journal of the American Society for Information Science, 41*(6), 433.

Mukherjee, D., Lim, W. M., Kumar, S., & Donthu, N. (2022). Guidelines for advancing theory and practice through bibliometric research. *Journal of Business Research, 148*, 101–115.

Mulder, I., & Koellner, T. (2011). Hardwiring green: How banks account for biodiversity risks and opportunities. *Journal of Sustainable Finance & Investment, 1*(2), 103–120.

Mundaca, L., & Heintze, J. N. (2024). Banking on ecosystem services. *Ecological Economics, 224*, Article 108284.

Nedopil, C. (2023). Integrating biodiversity into financial decision-making: Challenges and four principles. *Business Strategy and the Environment, 32*(4), 1619–1633.

Pandey, D. K., Hassan, M. K., Kumari, V., Zaied, Y. B., & Rai, V. K. (2024). Mapping the landscape of FinTech in banking and finance: A bibliometric review. *Research in International Business and Finance, 67*, Article 102116.

Pattnaik, D., Hassan, M. K., Kumar, S., & Paul, J. (2020). Trade credit research before and after the global financial crisis of 2008—A bibliometric overview. *Research in International Business and Finance, 54*, Article 101287.

Raimondi, R., Schwizer, P. G. M., Cosma, S., & Rimo, G. (2025). Credit risk assessment in the climate shadow: Evidence from white and grey literature. *Business Strategy and the Environment, 2025*, 1.

Randalls, S. (2009). Managing financial risks in urban *Managing financial risks: From global to local* (p. 209).

Sami, F., Faiz, U., & Khan, I. H. (2025). Sustainable finance: Pathways to a green economy. In *Adaptive strategies for green economy and sustainability policies* (pp. 425–442). IGI Global Scientific Publishing.

Stevenson, P. (2024). Shifting bank funding away from factory farming. In *Regenerative farming and sustainable diets* (pp. 203–212). Routledge.

Stolbov, M., Shchepeleva, M., & Parfenov, D. (2025). What is the relationship between biodiversity and the frequency of financial crises? Global Evidence. *Economics Letters, 250*, Article 112259.

Su, C. W., Mirza, N., Umar, M., Chang, T., & Albu, L. L. (2022). Resource extraction, greenhouse emissions, and banking performance. *Resources Policy, 79*, Article 103122.

Van Eck, N., & Waltman, L. (2010). Survey software: VOSviewer, a computer program for bibliometric mapping. *Scientometrics, 84*(2), 523–538.

van't Klooster, J., & Prodani, K. (2025). Planetary financial policy and the riskification of nature. *Review of International Political Economy, 32*(3), 643–667.

Vávrová, E. (2009). The current demand for the insurability of liability for damage to the natural environment. *Agricultural Economics, 55*, 33–39.

Zhu, Y., Prescott, G. W., Chu, P., & Carrasco, L. R. (2024). Glaring gaps in tools to estimate businesses' biodiversity impacts hinder alignment with the Kunming-Montreal global biodiversity framework. *Journal of Cleaner Production, 451*, 142079.

CHAPTER 4

Quantifying Biodiversity Exposure: Metrics, Methods, and Limitations

Abstract This chapter explores how biodiversity loss becomes a source of financial risk for banks. It outlines the main transmission channels through which ecosystem degradation affects credit quality, collateral value, liquidity, and operational resilience. The chapter then reviews the main tools for measuring biodiversity-related impacts and dependencies, distinguishing between sectoral (e.g., ENCORE, Sectoral Materiality Tool), geographical (WWF Biodiversity Risk Filter), and firm-level approaches (Corporate Biodiversity Footprint, biodiversity-adjusted ESG scores, and text-based metrics). Emerging models, such as Bio-Value-at-Risk (BioVaR), link ecological shocks to financial losses, thereby bridging the gap between environmental science and risk modelling. These methodologies align with the Taskforce on Nature-related Financial Disclosures (TNFD), supporting financial institutions in identifying, quantifying, and integrating nature-related risks into their strategies and disclosures. The chapter concludes by emphasising current limitations such as data scarcity, methodological heterogeneity, and weak links between ecological and financial metrics, while stressing the strategic role of robust biodiversity assessment in safeguarding financial stability and steering capital towards nature-positive outcomes.

Keywords Biodiversity finance · Biodiversity risk · Biodiversity metrics · Biodiversity footprint · Nature-related risks

G. Rimo and S. Cosma, *Biodiversity and Banks*, Palgrave Studies in Impact Finance, https://doi.org/10.1007/978-3-032-17631-8_4

4.1 Banks' Exposure to Biodiversity Loss

Serving as intermediaries between savers and borrowers, allocating capital across various sectors and geographical areas, banks are not only transmitters of financial shocks but also particularly vulnerable to risks that originate outside traditional financial domains. Among these, biodiversity loss has emerged as a critical, though still under-recognised, source of risk. Financial institutions, although only minimally reliant on natural capital themselves, are indirectly exposed to biodiversity risks through their lending and investment activities, which link them to the dependencies and impacts of non-financial firms (Kedward et al., 2021). Moreover, unlike non-financial corporations, whose exposure is often limited to their sectoral operations, banks tend to aggregate risks across industries and geographies, making them vulnerable not only to idiosyncratic but also to systemic shocks.

As previously seen and as highlighted by the Network for Greening the Financial System (NGFS), the erosion of ecosystems and natural capital has the potential to undermine firms' capacity to generate stable revenues, alter the value of collateral, and increase the probability of default (Boldrini et al., 2023). Financial institutions are often exposed to losses due to the declining performance of assets or economic activities that depend on biodiversity.

In parallel, regulatory changes, litigation, and investor activism aimed at reducing biodiversity impacts are reshaping entire industries, modifying the risk–return profile of banks' lending and investment portfolios. Governments' actions, technological advancements, legal proceedings, and changing consumer preferences aimed at reducing biodiversity loss and ecosystem degradation, largely driven by economic actors responsible for these impacts, can create transition risks. This is especially true when financial institutions are directly or indirectly exposed to those economic actors (NGFS-INSPIRE, 2021a, 2021b).

Furthermore, financial institutions are exposed to "lender liability", that is, the reputational risk associated with providing capital to environmentally harmful activities (Adeabah et al., 2023; Mulder & Koellner, 2011). In this regard, banks may incur reputational harm, legal repercussions, or direct financial losses when the firms they finance cause biodiversity loss. Historically, activism has always pressured the financial sector and its involvement in controversial industries (Cojoianu et al.,

2021; Emma & Jennifer, 2021). Organisations like *BankTrack* (www.banktrack.org) monitor and document how banks finance high-impact industries. In particular, BankTrack's "*Banks & Nature*" campaign focuses on exposing and reducing the role of banks in biodiversity loss. The organisation pressures financial institutions to stop financing activities that harm ecosystems and instead direct capital towards projects that protect and restore nature. It calls on banks to recognise the biodiversity crisis, align their strategies with the Global Biodiversity Framework, and adopt concrete transition plans with measurable targets. Finally, the campaign identifies "no-go" areas where bank financing should be excluded altogether, such as industrial livestock, deep-sea mining, and biomass combustion, as well as intact forests, free-flowing rivers, and indigenous territories where the right to free, prior, and informed consent must be respected.

4.1.1 Channels of Exposure

In other words, biodiversity loss reaches bank balance sheets through different pathways. Some are visible and sector-specific, others are hidden in supply chains, and still others operate at the level of the entire financial system. Mapping these channels is a necessary step before exploring how to measure and manage them. Indeed, to understand these dynamics, it is useful to distinguish between three main channels of exposure (Bolton et al., 2020; Calice et al., 2021):

- **Direct exposures** arise from loans and investments in biodiversity-dependent or biodiversity-impacting sectors. Agriculture, fisheries, forestry, tourism, and extractive industries are critical sectors whose performance relies heavily on ecosystem services such as pollination, soil fertility, and water availability. However, the operations of these industries often lead to biodiversity degradation. For banks, this situation results in increased credit and reputational risks. For instance, consider a coastal tourist area that thrives on healthy marine ecosystems, including clean water, intact beaches, coral reefs, and rich marine biodiversity, which attract visitors. When biodiversity is degraded due to factors such as coastal erosion, coral bleaching, pollution, and habitat loss, the area becomes less attractive, resulting in a decline in tourism. Consequently, local businesses, such as hotels, restaurants, beach resorts, and diving operators, experience a sharp drop in revenues and profit margins. As these businesses struggle, they may find it challenging to repay their loans and may request moratoria, causing an increase in non-performing loans (NPLs) within the lending bank's portfolio. This situation significantly elevates the bank's credit risk.

Furthermore, if the bank has funded projects in ecologically sensitive areas that contribute to erosion, pollution, or habitat destruction, it risks facing backlash from non-governmental organisations (NGOs), the media, and the public. Such scrutiny could lead to accusations of supporting environmentally harmful activities. As a result, the bank may encounter negative media coverage, public protests, and a loss of trust from customers and investors, all of which represent direct reputational risks.

- **Indirect exposures** are transmitted through clients' supply chains and capital markets. Even firms in apparently low-impact industries may rely on upstream suppliers that depend on biodiversity (for example, food retailers or textile companies). Banks also carry biodiversity-related risk through their trading and asset management activities, holding securities issued by firms with nature-related dependencies. These exposures are often diffuse and harder to measure, amplifying the problem of risk mispricing. For instance, a large international fast fashion chain appears to operate in a sector with a relatively low impact on biodiversity (retail, urban commerce). In reality, its upstream supply chain is deeply linked to vulnerable ecosystems. Cotton production requires substantial amounts of water, often in regions already experiencing water stress. Intensive cultivation leads to soil erosion and a decline in soil fertility. Pesticide use compromises local ecosystems and reduces biodiversity. If a severe drought or deterioration in soil quality affects these regions, cotton production declines significantly. The retailer may experience increased raw material costs, supply delays, lower sales, and margin compression. Although not a "high-impact company" it therefore suffers financial consequences from a biodiversity shock that affects its supply chain. The bank that finances the retailer (corporate credit) or holds its bonds or shares in its trading or asset management portfolio is exposed to an indirect but significant deterioration: credit risk (decrease in repayment capacity), market risk (loss of value of bonds/shares), operational, and reputational risk if allegations arise about the supply chain (deforestation, water exploitation, etc.). The exposure does not come from the "direct" customer but from its primary and secondary suppliers, often located in different countries and with highly heterogeneous environmental impacts. For this reason, the risk is widespread, opaque, and easily underestimated, contributing to the problem of mispricing nature-related risks.

- **Systemic exposures**, because of the complexity, interdependence, and interconnectedness within the financial system (Calice et al., 2021). Systemic exposures occur when biodiversity loss generates economy-wide shocks that simultaneously affect households, firms, and markets. For example, a sharp decline in pollinators (bees, bumblebees, and wild insects) drastically reduces agricultural productivity across a vast area. Farms produce less, generate lower revenues, and many are no longer able to repay bank loans. Farmers also reduce purchases from suppliers (seeds, logistics, packaging), which in turn leads to difficulties. Food industries must pay significantly more for raw materials or are unable to obtain them. Food prices rise. Consumers reduce their spending or experience a decline in real income. The simultaneous increase in NPLs for agricultural businesses, suppliers, and SMEs in the supply chain and the loss in the value of collateral (e.g., less productive agricultural land means lower collateral value) affect credit risk for banks; if many sectors require restructuring or moratoria, liquidity risks increase. Furthermore, regional banks, which are highly exposed to those sectors, may experience deterioration in their portfolios. However, due to interconnection among banks (including interbank lending, shared exposures, and participation in the same markets), the deterioration in one geographical area or sector can quickly spread to the rest of the financial system. Finally, considering banks' interconnections, shocks in one sector or region can spread rapidly through feedback loops, undermining financial stability.

Figure 4.1, developed by the NGFS, summarises the transmission mechanisms through which biodiversity loss can influence financial activities, financial institutions, and systems (NGFS-INSPIRE, 2021a, 2021b).

Biological diversity is at the top of the causal chain. Its degradation gives rise to physical and transition risks (defined in Chapter 2). These environmental pressures translate into economic disruptions, including capital destruction, higher volatility in raw material prices, operational cost increases, and the impairment of biodiversity-dependent services.

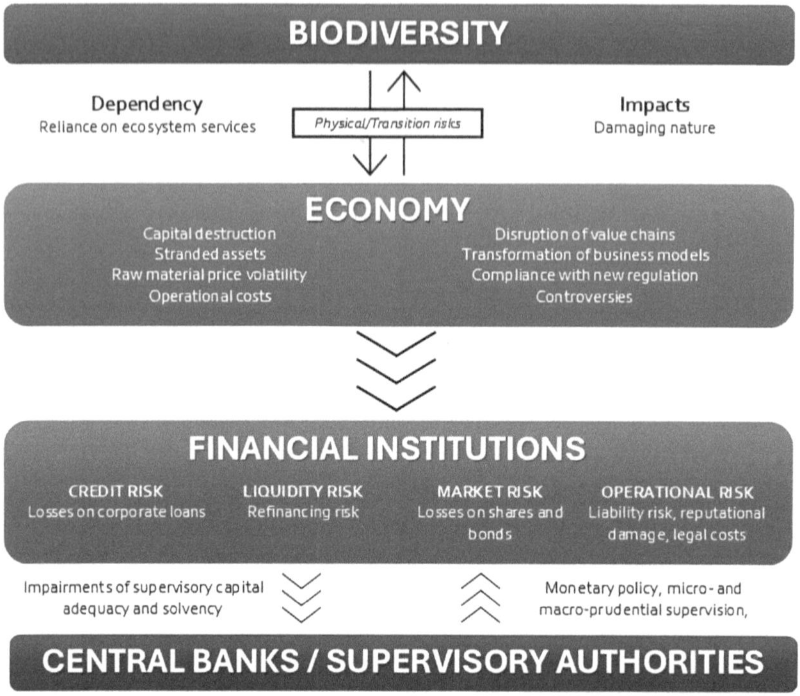

Fig. 4.1 The interaction between biodiversity loss, economic activities, and the financial system. *Source* Authors' elaboration on information from NGFS (2021)

Businesses may also face stranded assets or be forced to adapt through relocation and costly transformation of their production models.

These environmental shocks have significant implications for financial institutions, which are exposed through multiple, often interrelated risk channels. Indeed, these shocks may propagate into the financial system, testing the resilience of both banks and investors.

From a **credit risk** perspective, biodiversity degradation can impair borrowers' repayment capacity, particularly in sectors that rely directly on natural capital, such as agriculture, forestry, fisheries, tourism, and pharmaceuticals. As revenues decline, input costs rise, or regulatory constraints tighten, firms' cash flows become more volatile, and their solvency weakens, leading to higher probabilities of default (Adamolekun, 2024). Collateral linked to natural assets, such as farmland, fisheries, and forest concessions, may also depreciate in value, increasing banks' loss-given-default rates and reducing recovery rates in the event of borrower distress.

Liquidity risk may also arise as biodiversity loss affects the real economy and investor confidence. Firms experiencing biodiversity-related disruptions often draw on existing credit lines to compensate for cash-flow shortages, thereby creating unexpected funding demands on banks (Ahmad & Karpuz, 2024). At the same time, shifts in market sentiment can lead to rapid sell-offs of securities associated with biodiversity-sensitive activities, drying up market liquidity and widening bid–ask spreads. These dynamics can compel banks to liquidate positions under adverse conditions, resulting in liquidity mismatches and, in extreme cases, contagion effects across financial markets.

Banks and institutional investors may also be exposed to **market risk** due to the decline in corporate fundamentals in industries sensitive to biodiversity (Cosma et al., 2025). Bond spreads widen, credit ratings are downgraded, and equity prices decline as investors reassess the financial sustainability of the exposed firms. Banks and asset managers that hold these securities suffer mark-to-market losses as a result of these developments. Furthermore, sudden increases in market volatility can be triggered by policy shocks, legal disputes, or reputational crises related to biodiversity. This can increase Value-at-Risk and jeopardise the stability of portfolios that are concentrated in susceptible industries or regions.

Finally, biodiversity loss can increase **operational risk** and **reputational risk** through both physical and transition channels. Physical impacts, such as floods, droughts, soil erosion, or the collapse of local ecosystems, can disrupt banks' infrastructure, damage branch networks and data centres, and impair operational continuity. Transition-related impacts, by contrast, involve fines and legal risks associated with financing activities that harm biodiversity (Cosma et al., 2025). Financial institutions linked to environmentally harmful projects may face litigation, supervisory sanctions, or public backlash, all of which can lead to direct financial losses and erosion of franchise value.

Taken together, these dynamics demonstrate that biodiversity degradation is not merely an ecological concern but a source of traditional financial risks with the potential to affect banks' credit quality, asset valuations, liquidity positions, and operational resilience. Integrating nature-related factors into risk management frameworks is, therefore, essential to safeguard financial stability and align banking practices with the evolving prudential understanding of environmental risks.

The situation described above can also trigger a vicious circle as biodiversity loss leads to macroeconomic deterioration, financial risks increase, underscoring the urgent need to integrate nature-related financial risks into the core of supervisory and financial decision-making. Biodiversity loss can therefore affect the capital adequacy and solvency of financial companies. This reduces their ability to manage external shocks. Organisations that fail to integrate biodiversity into their risk management plans may be subject to government sanctions, increased capital requirements, or limited access to sustainable finance markets.

These exposures can accumulate and amplify at the system level, creating potential channels for systemic risk. For this reason, the diagram places central banks and financial supervisors at the bottom of the cascade, as they hold the mandate to preserve financial stability through monetary policy, *microprudential* oversight, and macroprudential interventions.

4.1.2 The Role of Banks in Addressing Biodiversity Loss

Banks can play a crucial role in combating biodiversity loss. Their ability to produce and evaluate high-quality information about the policies, strategies, and activities of financed entities is essential during the credit and investment screening phase. This role is enhanced by continuous monitoring of fund usage and managerial behaviour, as well as the contractual ability to terminate credit agreements if environmental or social conditions are violated. Together, these tools give banks significant power to guide businesses, public administrations, and households towards actions that protect ecosystems.

By integrating environmental criteria into credit-granting decisions, project finance, private equity, and investment activities (including shareholder activism), banks can directly influence corporate strategies. Systematic incorporation of biodiversity-related risks and opportunities into financing analyses, such as assessing projects' exposure to ecosystem degradation, deforestation, or unsustainable land use, allows for a reallocation of capital towards less impactful and more resilient production models.

The banking system also plays a vital role in promoting sustainable development and protecting natural capital. Banks can support economic activities that reduce pressure on ecosystems by financing and insuring renewable energy projects, the circular economy, and resilient water and urban infrastructure. This reduces physical and transition risks related to nature. They can also employ pricing schemes that incentivise biodiversity-friendly behaviour, such as loans with interest rates linked to environmental performance or reductions in ecological footprints. Additionally, banks can share data and expertise on environmental risks, which is essential for informing public policies and territorial planning. Moreover, adopting voluntary principles and standards, such as the Equator Principles, Principles for Responsible Banking (PRB), and Taskforce on Nature-related Financial Disclosures (TNFD), helps raise market expectations and encourages companies in sectors with high biodiversity impacts to improve their practices.

In doing so, the banking system not only mitigates its own financial risks but also becomes a key player in redirecting capital flows towards activities that support ecosystem conservation and the regeneration of natural capital. This allocative function, combined with its monitoring and enforcement capacity, positions banks as one of the main indirect economic policy instruments for combating biodiversity loss.

To effectively address biodiversity loss, banks should develop new risk management methods that include screening tools to assess the negative impacts of their investments and financing on biodiversity, as well as the potential spillover effects of biodiversity loss on traditional banking risks. Furthermore, to avert severe scenarios related to biodiversity loss, financial institutions and asset managers must undertake actions aimed at identifying resources for biodiversity conservation projects. By facilitating capital flows to sustainable economic activities, finance can serve as both a driver of biodiversity loss and a catalyst for its protection and restoration.

A net-zero and nature-positive future requires a shift in private capital flows from polluting companies and activities to those with a low environmental impact and a green approach. Financing activities that harm biodiversity, such as those related to fossil fuels, deforestation, habitat degradation, or pollution-intensive sectors, not only worsen ecological degradation but also expose financial institutions to increasing physical and transition risks. Conversely, aligning financial flows with biodiversity-positive objectives can help close the estimated annual biodiversity finance gap of US$598–824 billion (Deutz et al., 2020).

While the financial system cannot solve biodiversity loss on its own, it carries a critical responsibility in steering economic incentives and capital allocation (Cosma et al., 2023). This necessitates a rethinking of financial and prudential policy frameworks to enable the mobilisation of capital towards nature-positive outcomes, potentially through the removal of legal and institutional barriers and the promotion of financial innovations that support conservation and ecosystem restoration.

4.2 Tools and Metrics to Measure Biodiversity-Related Impacts and Dependencies

Having outlined how banks and financial institutions are exposed to biodiversity-related risks, the next step is to understand how these exposures can be measured and quantified. The translation of nature-related dependencies and impacts into financial metrics remains one of the most significant challenges (Giglio et al., 2023). Unlike climate risk, for which relatively standardised indicators, such as greenhouse gas emissions, are now widely available, biodiversity loss involves multiple dimensions, spatial heterogeneity, and complex feedback mechanisms that make measurement inherently difficult.

A growing ecosystem of tools and frameworks is currently emerging to link ecological data and financial risk assessment. These tools vary not only in methodology but also in their scale of analysis and whether they emphasise impacts (i.e., pressures exerted on ecosystems) or dependencies (i.e., reliance on services provided by ecosystems). We can distinguish these tools by their unit of analysis:

- **Sectoral tools** operate at the level of industries or value chains. They help banks to screen and map which sectors are most exposed to nature-related risks or most likely to impact biodiversity. These tools are particularly useful early in portfolio construction or sector allocation decisions, providing a first layer of risk filtering.
- **Geographic (spatial) tools** localise exposure by incorporating ecosystem data, environmental indicators, and spatial context. They identify hotspots of biodiversity risk in specific geographies, sites, or landscapes, thereby refining exposure assessments at the "sector + region" level. For banks with geographically diverse portfolios, these tools enable spatially aware risk analysis.
- **Firm-level tools (corporate metrics)** zoom in on the level of individual companies. They estimate a firm's direct and indirect impacts on biodiversity (e.g., habitat loss, species decline) or its dependency

on ecosystem services. These metrics can then be aggregated to form portfolio-level biodiversity exposure profiles or stress-test results.

Together, these instruments form the empirical foundation for integrating biodiversity considerations into credit analysis, portfolio management, and prudential supervision.

However, despite the progress made, these tools remain in an experimental phase and are mainly used for scientific purposes. Data gaps, methodological inconsistencies, and limited sectoral alignment constrain their comparability and use in decision-making. As regulators and market actors increasingly recognise the materiality of nature-related risks, improving the robustness and interoperability of biodiversity metrics has become a priority for both supervisors and financial institutions.

The following sections introduce and critically discuss the main tools currently available for assessing biodiversity-related impacts and dependencies, distinguishing their methodological foundations, units of analysis, and practical applications for banks.

4.2.1 Sectoral Tools

At the sectoral level, biodiversity-related financial risks are primarily assessed through tools that map the dependencies and impacts of entire industries on ecosystem services. By utilising these tools, financial institutions can pinpoint which areas of their portfolios are most vulnerable to biodiversity loss and, consequently, identify where lending and investment may be more susceptible to natural disasters. They are especially helpful in the early phases of sector allocation, portfolio screening, and strategic risk assessment, as they operate at a macro-sectoral scale.

Among the most established tools in this category are ENCORE (Exploring Natural Capital Opportunities, Risks, and Exposure) and the Sectoral Materiality Tool (SMT), jointly developed by the Taskforce on Nature-related Financial Disclosures (TNFD) and the Science-Based Targets Network (SBTN).

ENCORE is a collaboration among Global Canopy, the UNEP Finance Initiative, and the UN Environment Programme's World Conservation Monitoring Centre (UNEP-WCMC).[1] As reported on the official

[1] https://ENCOREnature.org/news/major-upgrade-for-ENCORE-launches-july-2024.

website, it was launched in 2018 to help financial institutions and companies understand how their activities depend on nature and impact it. It provides a structured approach to exploring potential nature-related risks by linking sectors of the economy with the ecosystem services they rely on and the pressures they exert on natural capital (Boldrini et al., 2023). At the heart of ENCORE lies a dual analytical pathway. The *dependency pathway* identifies how specific economic activities depend on ecosystem services such as water regulation, pollination, or soil fertility. The *impact pathway*, instead, traces how those same activities affect the natural environment through pressures such as land use, greenhouse gas emissions, or pollution, and how these translate into changes in ecosystems. Together, the two pathways provide an integrated view of the economy–nature interface, connecting business processes with biophysical systems (Table 4.1). In the latest updated version, economic activities are classified using the UN International Standard Industrial Classification (ISIC), which covers 271 activities worldwide. The mapping of ecosystem services follows the System of Environmental-Economic Accounting—Ecosystem Accounting (SEEA-EA), which is consistent with the TNFD and other corporate frameworks. The model also incorporates the Driver–Pressure–State–Impact–Response (DPSIR) logic to capture how human pressures translate into ecological change, and it distinguishes among 15 ecosystem types and eight ecosystem components (such as water, soils, atmosphere, and species).

Table 4.1 Dependency and impact pathways employed by ENCORE

Pathway	*Objective*	*Core analytical elements*
Dependency pathway	Identifies how economic activities rely on ecosystem services for their operations and value creation	Links between ISIC-classified economic activities and ecosystem services (as defined by the SEEA-EA framework)
Impact pathway	Assesses how economic activities exert pressures on ecosystems, potentially degrading natural capital	Follows the DPSIR framework (Driver–Pressure–State–Impact–Response), linking activities to pressures, which in turn drive mechanisms of change, ultimately affecting ecosystem components

Note This table describes the dual analytical pathway employed by ENCORE to assess the nature-related risks associated with individual economic sectors. *Source* Authors' elaboration

For each economic activity, ENCORE assigns a **materiality rating,** from *very low* to *very high,* indicating the relative importance of its dependencies and pressures. These ratings are derived from a combination of quantitative data (mainly from the Environmentally Extended Multi-Regional Input–Output database, EXIOBASE) and expert assessment. The tool, therefore, serves as a screening instrument, highlighting where the most material relationships between business and nature are likely to occur, rather than providing site-specific or company-level measures.

In practical terms, ENCORE allows users to select a sector (for example, agriculture or mining) and immediately visualise which ecosystem services it depends on most strongly, what pressures it exerts, and how these are distributed along its value chain. By linking activities with both their dependencies and impacts, ENCORE helps identify the sectors most exposed to biodiversity and ecosystem risks, providing a consistent basis for further analysis within frameworks such as the Taskforce on Nature-related Financial Disclosures (TNFD) or the Science-Based Targets for Nature (SBTN).

In short, ENCORE translates complex ecological relationships into a coherent, sector-level map of risks and dependencies, supporting financial and corporate actors in understanding how the stability of natural systems underpins economic activity (read Box 4.1).

Box 4.1—Example: How a bank can use ENCORE to assess sectoral exposure Consider a commercial bank with significant lending to the agricultural and food sector, particularly to companies involved in crop cultivation and food processing. To understand potential exposure to biodiversity and ecosystem risks, the bank uses ENCORE to screen its portfolio.

Step 1. Identify Relevant Activities

The risk team maps each client's main line of business to the corresponding **ISIC categories** used in ENCORE, e.g., *"Growing of cereals (except rice)"* and *"Processing and preserving of fruits and vegetables"*.

Step 2. Explore Dependencies

ENCORE shows that crop cultivation has *very high dependencies* on several ecosystem services, such as:

- Soil quality regulation (fertility and organic matter content);
- Water flow regulation and supply;

- Pollination services;
- Global and local climate regulation.

The tool's *materiality ratings* indicate that any degradation in these services would have *high functional and financial implications* for agricultural productivity.

Step 3. Examine Pressures

The same activities are linked to *very high pressures* on natural capital, such as:

- Land-use change and soil erosion,
- Emissions of nutrient pollutants to water and soil,
- Volume of freshwater use, and
- Greenhouse gas emissions.

Through ENCORE's **impact pathway**, these pressures are linked to changes in ecosystem components, such as soils, species populations, and water bodies, potentially undermining the same ecosystem services on which the sector relies.

Step 4. Interpret Implications for the Bank.

From this analysis, the bank concludes that its agricultural portfolio is exposed to **both dependency risk** (loss of ecosystem services could impair borrowers' production capacity) and **impact risk** (clients contribute to environmental degradation that may trigger stricter regulation or reputational damage). The institution therefore decides to:

- Prioritise the **agriculture sector** for a deeper, spatially explicit assessment
- Engage with key clients to discuss **water management and soil conservation practices**;
- Integrate ENCORE's *Very High* materiality ratings into its **sectoral credit policy**, requiring enhanced due diligence for new loans.

Result:

Using ENCORE, the bank has not only screened its exposure at the **sectoral level** but also identified **specific ecosystem dependencies and pressures** that can inform future risk management, lending decisions, and sustainability reporting.

The Sectoral Materiality Tool (SMT) extends this logic by systematically mapping the materiality of nature-related risks and opportunities across industries (Coqueret et al., 2025; Li et al., 2025). Developed by TNFD and the Science-Based Targets Network (SBTN), the SMT assesses each sector's level of dependency and impact across four dimensions of natural capital: land, ocean, freshwater, and atmosphere, and four subcategories of biodiversity-related risk: dependencies, impacts, risks, and opportunities. The tool builds on the scientific work of the Science-Based Targets Network and on the TNFD's LEAP approach (Locate, Evaluate, Assess, Prepare), translating ecological science into a decision-useful taxonomy for business and finance.

However, like ENCORE, the SMT is limited by its sectoral aggregation and by the heterogeneity of data sources used to evaluate each dimension of materiality. Its assessments provide indicative rather than quantitative results, meaning that it should be used as a screening and prioritisation tool, not as a measure of actual financial exposure. In practice, sectoral tools provide the conceptual and diagnostic foundation for more granular analyses at geographic and firm levels, which can capture the spatial variability and company-specific drivers of biodiversity risk.

4.2.2 Geographical Tools

While sectoral tools provide a macro-level understanding of which industries are most exposed to or responsible for biodiversity loss, they do not capture the spatial dimension of these risks. Biodiversity impacts and dependencies are inherently local: the same activity can have very different consequences depending on where it takes place, given the variability of ecosystems, species richness, and environmental regulations (Becker et al., 2025). Moreover, the academic literature suggests that considering country-level variables is crucial in assessing environmental risks, as well as in studying the role of space and place in shaping the transformation of firms and industries (Dicken & Malmberg, 2001). These studies suggest that firms derive some of their characteristics from specific territories and places and respond to the particularities of these locations (Schoenberger, 2002). In addition, according to the Broken Windows Theory of Urban Decline, negative environmental phenomena such as urban pollution can reduce the overall attention of firms and individuals to environmental issues (Tran et al., 2023).

Consequently, geographical tools have been developed to integrate location-specific ecological data into financial risk assessment, allowing users/banks to identify biodiversity "hotspots" and evaluate their exposure at the regional, national, or project level. Among the tools that assess biodiversity-related risks at a geographical or landscape level, the WWF Biodiversity Risk Filter (BRF) stands out as one of the most comprehensive and operationally mature.[2] Developed by the World Wide Fund for Nature (WWF), the BRF provides a spatially explicit assessment of biodiversity-related risks across terrestrial, freshwater, and marine ecosystems. It is designed to help companies and financial institutions identify, compare, and prioritise geographic areas where business activities are most exposed to biodiversity loss and ecosystem degradation.

The BRF operates within the broader WWF Risk Filter Suite, which also includes the Water Risk Filter. Both tools share a unified data infrastructure and can be accessed through the same digital platform, allowing users to conduct integrated biodiversity and water risk screenings. While ENCORE focuses on identifying sectoral dependencies and impacts on natural capital, the BRF complements it by providing a geospatial lens that evaluates where such dependencies and impacts are likely to materialise most critically.

At its core, the BRF uses a hierarchical and modular structure that combines information on the industry profile of an activity with the ecological characteristics of its location (WWF, 2024). The assessment framework is organised around two main dimensions of biodiversity risk:

- Physical risks, associated with the loss of ecosystem services on which firms depend (e.g., water supply, pollination, soil fertility);
- Reputational risks, related to the negative perception, social opposition, or regulatory scrutiny that may arise from operating in or near ecologically sensitive areas.

The risk assessment is based on over 70 global datasets compiled from peer-reviewed, institutional, and remote-sensing sources, including metrics of biodiversity state, ecosystem integrity, land-use change, and species importance. These datasets are aggregated into 33 indicators, grouped under eight risk categories (five physical and three reputational). Each indicator is assigned a risk score on a five-point scale ranging from *very low* to *very high*, allowing comparability across locations and sectors.

[2] https://riskfilter.org/biodiversity/home.

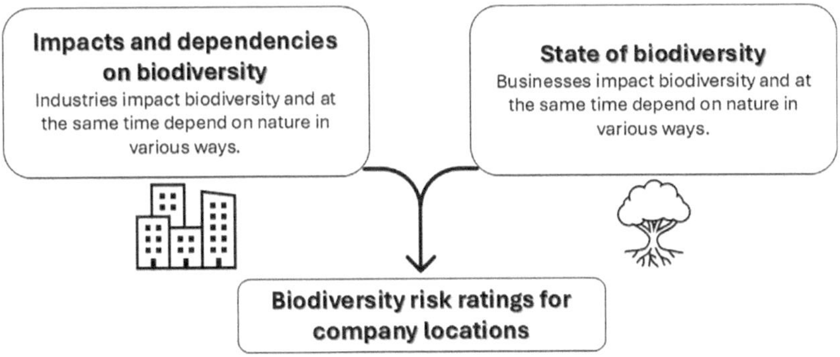

Fig. 4.2 Biodiversity-related risk assessment as a combination of the location of corporate activities and the importance and state of biodiversity integrity. *Source* Authors' elaboration on information provided by WWF (2024)

Methodologically, the BRF combines two types of information (Fig. 4.2):

1. **Industry materiality scores**, which capture the typical level of dependency and impact of each sector on biodiversity (derived from WWF's sectoral materiality matrix);
2. **Location-based assessments**, which measure the biodiversity importance and degradation level of the specific landscape or seascape where the activity is situated.

For each indicator, the final "scape risk score" is obtained as the average of the industry and location components, with higher values indicating greater exposure. The aggregation across indicators uses a *percentile-based approach* that emphasises high-risk factors rather than averaging them out, ensuring that a few severe local issues (e.g., deforestation or soil erosion) are not masked by otherwise moderate values.

The tool is accessed through an online Access module, where users can upload and manage their corporate or financial portfolios by georeferencing sites, projects, or assets. For each site, the BRF automatically retrieves relevant spatial data and generates both quantitative scores and visual outputs—maps, spider diagrams, and heatmaps—that highlight "biodiversity risk hotspots". These visualisations enable users to pinpoint areas where ecosystem degradation, conservation importance, and industrial sensitivity overlap, and to export data for further integration into internal risk management or disclosure processes.

Importantly, the BRF is conceived as a screening tool, not a local ecological impact assessment. It identifies where biodiversity-related risks are most likely to occur but does not quantify their financial magnitude nor account for firm-specific mitigation measures. Therefore, it is particularly suited for portfolio-level analysis, scenario screening, and the early stages of risk management—especially for banks and investors seeking to align with the Taskforce on Nature-related Financial Disclosures (TNFD) frameworks.

4.2.3 Firm-Level Tools

Firm-level tools represent the most granular category of biodiversity risk assessment instruments, focusing on the direct and indirect impacts and dependencies of individual companies. While sectoral and geographical tools help identify where biodiversity-related risks are concentrated, firm-level metrics enable a more precise evaluation of how corporate activities translate into measurable ecological pressures and how these, in turn, affect financial performance. Such tools are crucial for banks and investors seeking to integrate biodiversity considerations into credit analysis, portfolio monitoring, and engagement strategies.

Corporate Biodiversity Footprint

Among the most advanced tools developed to assess biodiversity impacts at the firm level, the Corporate Biodiversity Footprint (CBF), proposed by Iceberg Data Lab (IDL), represents a science-based and comprehensive approach that quantifies the annual biodiversity footprint of corporations, financial institutions, and sovereign issuers. The CBF translates the complex interactions between economic activities and ecosystems into a single, comparable metric, expressed in square kilometres of Mean Species Abundance (km^2 MSA). This unit measures the extent of biodiversity loss equivalent to the degradation of a given area of pristine habitat, thus providing an intuitive and standardised expression of ecological impact across sectors and geographies (Iceberg Data Lab, 2023).

At its core, the CBF methodology builds upon the Mean Species Abundance (MSA) indicator developed within the GLOBIO model by the *Netherlands Environmental Assessment Agency (PBL)*. The MSA reflects the average relative abundance of native species in a disturbed ecosystem compared to its natural, undisturbed baseline. A value of 100% indicates a pristine state of biodiversity, while a value approaching 0% corresponds to complete ecosystem degradation (Fig. 4.3). The GLOBIO framework quantifies this relationship through empirically derived damage functions that link anthropogenic pressures such as land-use change, pollution, and climate change to biodiversity loss across major terrestrial biomes.

The CBF adopts a life-cycle and value-chain perspective, in line with the Organisation Environmental Footprint (OEF) and Life Cycle Assessment (LCA) methodologies endorsed by the European Commission (Iceberg Data Lab, 2023). Each company's activities are modelled through IDL's proprietary input–output system (*Wunderpus*), which maps financial flows and supply-chain dependencies by NACE sector and incorporates sustainability attributes (such as environmental certifications and management practices). This process allows the estimation of all relevant pressures on biodiversity, which are then converted into MSA losses through scientifically validated pressure–impact functions.

In summary, the CBF methodology transforms a company's financial and operational data into a quantified biodiversity footprint through a chain of environmental pressure modelling and scientifically validated

Fig. 4.3 Photographic illustration of MSA variation for forest and grassland ecosystems. Authors' elaboration based on information from Iceberg Data Lab (2023)

damage functions. The methodology explicitly considers four main categories of pressure:

1. **Land-use change**, encompassing occupational, transformational, incremental, fragmentation, and encroachment effects that alter or fragment natural habitats;
2. **Air pollution**, capturing acidification and eutrophication processes caused by nitrogen and sulphur emissions;
3. **Water pollution**, including freshwater ecotoxicity from chemical compounds and marine plastic entanglement;
4. **Climate change**, reflecting the impact of greenhouse gas emissions on species distribution and ecosystem resilience.

All pressures are evaluated across Scope 1, 2, and 3 boundaries, in accordance with the Greenhouse Gas (GHG) Protocol, thereby covering both direct and indirect impacts along upstream and downstream value chains. The resulting biodiversity losses are aggregated to yield an absolute footprint (expressed in km^2·MSA) and relative indicators, such as biodiversity loss per million euros of turnover, capital employed, or enterprise value.

To enhance interpretability for investors and financial actors, IDL developed a derivative scoring system, the CBF Score, which ranks firms from 1 (least harmful) to 6 (most harmful), either within a specific sector (*best-in-class*) or across the overall universe (*best-in-universe*). A trend indicator complements this score by tracking changes in performance over time. These scores facilitate portfolio-level biodiversity analysis, integration into ESG frameworks, and alignment with EU disclosure regulations, including the Sustainable Finance Disclosure Regulation (SFDR) and the Corporate Sustainability Reporting Directive (CSRD).

Beyond its measurement capacity, the CBF provides a foundation for a "double materiality" approach, as it can be coupled with IDL's Dependency Score, which assesses firms' exposure to biodiversity-related dependencies (e.g., ecosystem services essential to production). Together, these indicators enable financial institutions to quantify both the impact on nature and their dependence on nature, aligning with the frameworks promoted by the Taskforce on Nature-related Financial Disclosures (TNFD) and the EU taxonomy (read Box 4.2).

Box 4.2—Using the Corporate Biodiversity Footprint (CBF) for portfolio screening: The case of HSBC and the Euronext ESG Biodiversity-Screened Index An illustrative example of how the Corporate Biodiversity Footprint (CBF) can be applied in financial practice is provided by *HSBC's Euronext ESG Biodiversity-Screened Index*, launched in 2021 as the world's first broad-based equity index explicitly integrating biodiversity considerations. The initiative demonstrates how biodiversity impact metrics can be incorporated into portfolio construction and investment decision-making (Finance for Biodiversity Foundation, 2024).

The index was designed to enable investors to consider the impact on natural capital when allocating capital and to provide a benchmark for identifying which companies to include or exclude based on their biodiversity performance. Starting from the Euronext World Index (comprising approximately 1500 large- and mid-cap companies), HSBC and Euronext applied a four-step screening methodology:

1. **Financial screening**—exclusion of small companies to ensure investability and liquidity;
2. **SRI screening**—removal of firms involved in controversial activities such as weapons, tobacco, or coal;
3. **ESG screening**—exclusion of companies with high overall ESG risk scores;
4. **Biodiversity screening**—exclusion of the 33% of companies with the highest biodiversity impact *within each sector*, using the **Corporate Biodiversity Footprint (CBF)** methodology.

For the final step, the CBF was chosen because it captures companies' full value chains and expresses biodiversity impact in a robust ecological unit, Mean Species Abundance (MSA), standardised per unit of investment (*km^2·MSA per million euros invested*). The biodiversity screening excluded the top-impact firms in each sector, retaining the 500 largest remaining companies to compose the Euronext ESG World Biodiversity-Screened Index.

The index is rebalanced quarterly, with biodiversity scores updated annually, reflecting the latest data from Iceberg Data Lab and CDC Biodiversité. The results are notable: compared with the original Euronext World Index, the biodiversity-screened version shows a 53% lower weighted CBF intensity, a 53% lower weighted GHG intensity, and a 14% lower ESG risk score. Importantly, the index's economic and financial performance remains broadly in line with the benchmark, demonstrating that biodiversity integration does not necessarily entail a trade-off with returns.

This case illustrates how biodiversity impact metrics can evolve from scientific and reporting tools into investment decision instruments, supporting both risk reduction and positive environmental alignment. The HSBC example underscores the potential for the CBF to serve as a quantitative basis for portfolio screening, thematic index design, and the development of biodiversity-linked financial products.

Despite its comprehensiveness, the CBF still faces methodological challenges, notably limited coverage of marine biodiversity, dependence on corporate data availability, and the need for further development of modules addressing invasive species, resource consumption, and water stress. Nevertheless, it remains one of the most scientifically robust and operationally scalable biodiversity accounting tools currently available, offering a harmonised framework through which firm-level impacts can be integrated into financial analysis, risk assessment, and sustainable investment strategies.

Biodiversity-Adjusted ESG Scores

A second family of firm-level approaches consists of ***biodiversity-adjusted ESG scores***, which integrate biodiversity indicators into traditional environmental, social, and governance (ESG) rating frameworks. Providers such as Refinitiv (now LSEG Data & Analytics), MSCI, and Moody's ESG Solutions have progressively integrated biodiversity-related dimensions, ranging from land-use intensity and natural capital policies to controversies surrounding deforestation or ecosystem management, into their environmental pillars. These enriched ESG datasets enable banks and investors to preliminarily screen companies based on their biodiversity performance, even when firm-level ecological data remain incomplete or unavailable.

However, the underlying construction of these indicators differs markedly across data providers. In a comparative study, Clément and Grenon (2025) mapped the biodiversity metrics offered by MSCI and LSEG against the ENCORE framework to evaluate their coverage of biodiversity-relevant environmental factors. The authors found that MSCI's dataset covers approximately 35% of ENCORE factors, while LSEG's (formerly Refinitiv) coverage reaches only 24%, with both providers offering extensive data on climate-related variables such as greenhouse gas emissions but showing substantial gaps for soil pollutants, water pollutants, and non-GHG air pollutants—all key determinants of biodiversity loss.

MSCI's approach combines modelled estimates and normalisation techniques to fill data gaps across industries and geographies. Its "Biodiversity & Land Use" key issue, one of 33 ESG Key Issues within MSCI's Natural Capital theme, evaluates a firm's *exposure* to biodiversity-related risks and its *management capacity* to mitigate them (Xin et al., 2025). Exposure scores take into account the firm's business model and geographic footprint, while management scores evaluate the presence of land-use policies, ecosystem restoration initiatives, and environmental governance structures (MSCI, 2023). MSCI also utilises geospatial overlays to identify corporate operations in biodiversity-sensitive areas and tracks indicators related to deforestation screening, freshwater dependency, and waste management. These metrics are predominantly derived from company reports, complemented by modelled data to ensure cross-sector comparability.

In contrast, LSEG (Refinitiv) employs a disclosure-based methodology (Clément & Grenon, 2025). Its ESG database, which includes over 400 indicators, relies exclusively on *reported data* rather than estimates. Biodiversity-related measures within Refinitiv's framework are therefore limited, focusing primarily on whether a company *discloses* information about its impact on biodiversity or the actions it takes to mitigate such impact. For instance, one key variable, *biodiversity impact reduction*, is a binary indicator equal to 1 if a firm reports activities addressing biodiversity impacts and 0 otherwise. This design ensures empirical consistency but results in sparse coverage and low granularity, particularly for firms and sectors that do not yet report on biodiversity issues.

Biodiversity-adjusted ESG scores should be interpreted as proxy indicators, useful for preliminary screening, engagement, and investor awareness, but insufficient for quantitative risk modelling or asset pricing. While the integration of biodiversity metrics into ESG ratings marks a significant step towards mainstreaming nature-related considerations in finance, these indicators remain qualitative and heterogeneous. They primarily reflect firms' transparency and governance responses rather than measurable ecological performance. Thus, biodiversity-adjusted ESG scores serve as an initial gateway for financial institutions to engage with nature-related risks, but their interpretive limits must be clearly acknowledged in any rigorous risk management or valuation framework.

Text-Based Biodiversity Exposure Metrics

A rapidly growing stream of research leverages text analysis and machine learning to identify firms' exposures to biodiversity risk from unstructured textual data such as corporate reports, financial filings, and media coverage. These approaches combine Natural Language Processing (NLP) techniques with supervised or transformer-based models to detect, classify, and quantify biodiversity-related content across large corpora of financial documents.

A leading example is the framework developed by Giglio et al. (2023), who construct several text-based biodiversity exposure indicators at both the aggregate and firm level. At the aggregate level, they derive a *Biodiversity News Index* from *New York Times* articles between 2010 and 2023. The authors build a "Biodiversity Dictionary" of 100 keywords semantically linked to biodiversity (identified through *word2vec*) and apply BERT (Bidirectional Encoder Representations from Transformers) to classify each biodiversity-related sentence as positive, negative, or neutral. The resulting index captures shifts in public attention and sentiment towards biodiversity loss and regulatory developments, functioning as a high-frequency proxy for *news-based biodiversity risk.*

At the firm level, the authors extract biodiversity-related information from 10-K filings using the same text-mining approach. They construct three indicators:

1. 10 K-Biodiversity-Count Score, measuring whether firms discuss biodiversity issues;
2. 10 K-Biodiversity-Negative Score, quantifying the net negative sentiment of those disclosures; and
3. 10 K-Biodiversity-Regulation Score, identifying references to regulatory or legal biodiversity risks.

These text-derived variables reveal that biodiversity mentions, especially those linked to regulation, have risen steadily over the past two decades, and that exposures are highly heterogeneous across industries (highest in energy, utilities, real estate, food, and pharmaceuticals).

For financial institutions, such text-based biodiversity metrics offer a scalable and complementary solution to bridge current data gaps. They can be used to map biodiversity-related exposures across loan or investment portfolios, detect emerging physical and transition risks, and support engagement or screening strategies. However, their accuracy depends on model architecture, keyword selection, and contextual interpretation, and they remain indirect proxies compared to quantitative impact measures such as the Corporate Biodiversity Footprint (CBF).

In summary, text-based approaches extend biodiversity risk assessment beyond quantitative footprints, enabling the detection of risk perceptions, disclosure intensity, and market sentiment in real time. When combined with firm-level and sectoral indicators such as the CBF or biodiversity-adjusted ESG scores, they provide a more comprehensive view of how nature-related risks materialise and are reflected in financial data.

Scenario-Based and Geospatial Quantitative Approaches: The Bio-Value-at-Risk (BioVaR) Framework (Posth et al., 2024)
Recent methodological advances are moving beyond descriptive biodiversity metrics towards quantitative, risk-based approaches that translate ecological impacts into financial loss distributions. A pioneering example is the *Bio-Value-at-Risk (BioVaR)* framework proposed by Posth et al. (2024), which combines geospatial biodiversity data with financial risk modelling. The BioVaR introduces a bottom-up, scenario-based methodology to quantify how biodiversity loss at specific corporate sites can affect a firm's balance sheet and, by aggregation, the risk profile of financial portfolios.

The approach proposed by Posth et al. (2024) integrates high-resolution geospatial information, derived from remote-sensing observations and global biodiversity variables, with asset-level operational data to estimate local biodiversity impacts attributable to each facility. These localised impacts are then converted into *site-specific financial losses* through empirically calibrated cost functions reflecting factors such as restoration expenses, operational dependencies, or regulatory penalties. The resulting monetary estimates are integrated into firms' risk management frameworks and financial statements, enabling the calculation of distribution-based risk measures, such as Expected Loss, Value-at-Risk, or Expected Shortfall, under biodiversity-related scenarios.

Conceptually, the BioVaR framework bridges environmental science and quantitative finance by transforming biodiversity losses, typically measured in ecological units such as Mean Species Abundance or Potentially Disappeared Fraction, into standardised financial metrics. This enables portfolio managers to assess biodiversity risk in a manner analogous to market or credit risk and to construct biodiversity-consistent stress-testing exercises. As such, BioVaR represents a new generation of *spatial-finance tools*, complementing corporate biodiversity footprints and ESG-based screening by offering a consistent, scenario-driven valuation of nature-related risks across companies and geographies.

4.2.4 From Measurement to Management: Applying Biodiversity Metrics in Financial Practice

The measurement approaches discussed above represent some of the tools through which financial institutions can translate complex ecological dynamics into decision-useful information. Yet, the ultimate value of these methodologies lies in their capacity to support a structured process of identifying, assessing, and managing nature-related risks and opportunities. In this regard, the Taskforce on Nature-related Financial Disclosures (TNFD) provides a unifying framework for integrating biodiversity and broader nature-related considerations into financial decision-making, risk management, and corporate disclosure.

Established in 2021, the TNFD aims to provide financial institutions and corporates with a consistent approach to assess and disclose their nature-related dependencies, impacts, risks, and opportunities. Its structure mirrors that of the Task Force on Climate-related Financial Disclosures (TCFD) but expands its scope beyond climate to encompass all realms of nature, i.e., land, ocean, freshwater, and atmosphere. The TNFD's framework revolves around four disclosure pillars: *governance*, *strategy*, *risk and impact management*, and *metrics and targets*, which guide entities in reporting how nature-related factors are integrated into corporate and financial practices (TNFD, 2023). The Taskforce has then developed an integrated approach to help corporates and financial institutions assess and manage nature-related issues. This approach is called LEAP, a stepwise analytical process designed to help organisations integrate nature-related considerations into their decision-making. LEAP stands for Locate, Evaluate, Assess, and Prepare, each phase corresponding to a distinct stage in understanding and managing exposure to nature-related issues:

- **Locate**: Identify the interfaces between business activities (or financial portfolios) and nature, determining where dependencies and impacts occur across geographies, sectors, and value chains.
- **Evaluate**: Quantify the dependencies and impacts identified in the previous phase, using appropriate indicators, models, or metrics.
- **Assess**: Analyse the financial materiality of nature-related risks and opportunities, considering exposure, sensitivity, and potential responses.
- **Prepare**: Integrate the findings into governance, strategy, and disclosure processes, aligning internal risk management and external reporting with the TNFD recommendations.

The different tools reviewed in this chapter can be meaningfully connected to the four phases of the LEAP process, as largely highlighted by the Finance for Biodiversity Foundation (2024) in their guide "*Biodiversity measurement approaches: A practitioner's guide for financial institutions*". These tools provide banks and financial institutions with a coherent pathway from identifying biodiversity exposure to integrating results into strategic and disclosure frameworks. Rather than being isolated instruments, these methodologies form a continuum that supports decision-making at different stages of maturity and data availability.

At the **Locate** stage, sectoral and spatial tools play a crucial role. ENCORE and the Sectoral Materiality Tool (SMT) allow users to identify which economic activities are most dependent on or exert the strongest pressures on nature, helping to prioritise the sectors where biodiversity risks are likely to be most material. The WWF Biodiversity Risk Filter (BRF) complements this analysis with a geographic perspective, enabling institutions to map where these pressures occur and to identify biodiversity-sensitive locations within their portfolios. Together, these tools help banks form an initial picture of their nature-related exposure, both across industries and across space.

In the **Evaluate** phase, attention shifts from broad exposure mapping to quantification. The Corporate Biodiversity Footprint (CBF) developed by Iceberg Data Lab represents one of the most advanced methods for translating corporate activities into measurable ecological impacts, expressed in units of Mean Species Abundance (MSA). At the same time, biodiversity-adjusted ESG indicators and text-based biodiversity exposure measures based on natural-language processing provide complementary ways to evaluate biodiversity performance or disclosure intensity when quantitative data are scarce.

The **Assess** phase involves linking ecological outcomes to financial materiality. Both the BRF and the CBF contribute to this step, offering insights into how biodiversity loss can translate into credit, market, or reputational risks. The emerging Bio-Value-at-Risk (BioVaR) framework goes further, transforming biodiversity degradation into potential financial losses through scenario-based and spatial-finance models. These tools help financial institutions understand the pathways through which biodiversity decline affects balance-sheet resilience, lending portfolios, and investment performance.

Finally, during the **Prepare** phase, institutions can utilise the evidence generated through previous steps to inform strategic actions and disclosures. Portfolio-level applications of the CBF, such as the Euronext ESG Biodiversity-Screened Index (see Box 4.2) exemplify how biodiversity metrics can inform capital allocation and thematic investment products. Scenario-based tools like BioVaR can, in turn, support the definition of risk appetites, internal targets, and external reporting aligned with evolving market and regulatory expectations.

Overall, these approaches provide a structured progression from identifying and quantifying biodiversity dependencies and impacts to assessing their financial relevance and integrating them into strategic management. By aligning with the logic of the LEAP process, they enable banks and financial institutions to move beyond qualitative awareness and towards a systematic, transparent, and science-based management of nature-related risks and opportunities.

4.3 Challenges and Limitations of Biodiversity Measurement Approaches

Although the field of measuring biodiversity has made significant advances recently, it remains challenging to convert ecological complexity into financially meaningful metrics that can inform decision-making. While the sectoral, geographic, and firm-level tool families discussed in this chapter provide insightful information about potential impacts and dependencies, they all have structural limitations that should be carefully considered when using them in financial practice.

First, there is a persistent data and coverage gap. Most tools evaluate potential impacts and dependencies instead of actual, measurable changes in biodiversity. As emphasised by the Finance for Biodiversity Foundation (2024), modelled impacts may diverge significantly from real outcomes, as they depend on assumptions, sectoral averages, and incomplete datasets. For example, ENCORE relies on sector-level materiality ratings rather than company-specific data, whereas footprinting approaches, such as the Corporate Biodiversity Footprint (CBF), depend on modelled relationships between pressures and biodiversity responses. Consequently, the precision of these tools is limited by the availability and quality of corporate disclosures and geospatial data.

Second, users must be aware that the results of modelled tools should be employed only for appropriate purposes, such as estimating portfolio-level exposure, monitoring and screening biodiversity risks, guiding engagement activities, or training portfolio managers, rather than for stock selection or quantitative valuation of individual companies. As the *Practitioner's Guide* suggests, modelled results are most valuable for identifying hotspots, understanding the relative contributions of different sectors or drivers, and prioritising further due diligence or data collection efforts, rather than for fine-grained risk pricing or capital allocation (Finance for Biodiversity Foundation, 2024).

Third, the limited scope and uneven coverage of biodiversity drivers constrain the representativeness of current approaches. Many tools under-represent the impacts on marine ecosystems (for example, the Corporate Biodiversity Footprint) and do not yet include pressures such as invasive alien species, resource extraction beyond water use, or downstream impacts across value chains.

Fourth, significant methodological heterogeneity and lack of interoperability persist among tools. As stated by Roeder and Utz (2025), "*while providers more or less agree on which firm operations contribute to a loss in biodiversity, there is a fundamental disagreement about the underlying methods and data to measure them*". Sectoral instruments rely on qualitative materiality scales, geographical tools combine spatial indicators through expert weighting, and firm-level methods use life-cycle or scenario-based modelling. These conceptual differences complicate comparability and integration into unified risk frameworks. Currently, no single methodology can provide a comprehensive, consistent, and cross-sectoral measure of biodiversity risk.

Fifth, linking ecological indicators to financial materiality remains a strong challenge. Converting ecological units, such as Mean Species Abundance (MSA), into financial risk parameters requires novel modelling and empirical validation. While frameworks like BioVaR represent promising steps, they still rely on assumptions about the propagation of ecological shocks into balance-sheet losses (Posth et al., 2024). The causal chain from biodiversity degradation to financial outcomes is inherently long, non-linear, and context-specific.

Financial institutions must therefore build internal capacity to interpret, contextualise, and combine these data responsibly. Without transparent communication of assumptions, scope boundaries, and uncertainties, there is a risk of misusing biodiversity metrics or overstating their precision.

References

Adamolekun, G. (2024). Firm biodiversity risk, climate vulnerabilities, and bankruptcy risk. *Journal of International Financial Markets, Institutions and Money, 97*, Article 102075.

Adeabah, D., Andoh, C., Asongu, S., & Gemegah, A. (2023). Reputational risks in banks: A review of research themes, frameworks, methods, and future research directions. *Journal of Economic Surveys, 37*(2), 321–350.

Ahmad, M. F., & Karpuz, A. (2024). Beyond climate change risk: Biodiversity and corporate cash holdings. *Economics Letters, 236*, Article 111608.

Becker, A., Di Girolamo, F. E., & Rho, C. (2025). Loan pricing and biodiversity exposure: Nature-related spillovers to the financial sector. *Research in International Business and Finance, 75*, Article 102724.

Boldrini, S., Ceglar, A., Lelli, C., Parisi, L., & Heemskerk, I. (2023). *Living in a world of disappearing nature: Physical risk and the implications for financial stability*. ECB occasional paper (2023/333).

Bolton, P., Despres, M., Da Silva, L. A. P., Samama, F., & Svartzman, R. (2020). The green swan. *BIS Books*.

Calice, P., Kalan, F. D., & Miguel, F. (2021). Nature-related financial risks in Brazil. *Policy research working papers*.

Clément, A., & Grenon, G. (2025). Mapping biodiversity risk metrics in finance: A comparative analysis of MSCI and LSEG using the ENCORE framework. *Journal of Sustainable Finance and Accounting, 2025*, 100020.

Cojoianu, T. F., Ascui, F., Clark, G. L., Hoepner, A. G., & Wójcik, D. (2021). Does the fossil fuel divestment movement impact new oil and gas fundraising? *Journal of Economic Geography, 21*(1), 141–164.

Coqueret, G., Giroux, T., & Zerbib, O. D. (2025). The biodiversity premium. *Ecological Economics, 228*, Article 108435.

Cosma, S., Cosma, S., Pennetta, D., & Rimo, G. (2025). Does biodiversity matter for firm value? *Journal of International Financial Markets, Institutions and Money, 105*, Article 102240.

Cosma, S., Rimo, G., & Cosma, S. (2023). Conservation finance: What are we not doing? A review and research agenda. *Journal of Environmental Management, 336*, Article 117649.

Deutz, A., Heal, G. M., Niu, R., Swanson, E., Townsend, T., Zhi L., Delmar, A., Meghji, A., Sethi, S. A., & Tobin-de la Puenta, J. (2020). Financing nature: Closing the global biodiversity financing gap. *The Paulson Institute, The Nature Conservancy and Cornell Atkinson Center for Sustainability.*

Dicken, P., & Malmberg, A. (2001). Firms in territories: A relational perspective. *Economic Geography, 77*(4), 345–363.

Emma, G. M., & Jennifer, M. F. (2021). Is SDG reporting substantial or symbolic? An examination of controversial and environmentally sensitive industries. *Journal of Cleaner Production, 298*, Article 126781.

Finance for Biodiversity Foundation. (2024). *Biodiversity measurement approaches A practitioner's guide for financial institutions*. Available at: https://www.financeforbiodiversity.org/publications/

Giglio, S., Kuchler, T., Stroebel, J., & Zeng, X. (2023). *Biodiversity risk* (No. w31137). National bureau of economic research.

Iceberg Data Lab. (2023). *Corporate biodiversity footprint-methodological guide*. Available at: https://www.icebergdatalab.com/wp-content/uploads/2023/04/CBF_client_methodological_guide_March_23.pdf

Kedward, K., Ryan-Collins, J., & Chenet, H. (2021). *Understanding the financial risks of nature loss: Exploring policy options for financial authorities*. SUERF: The European Money and Finance Forum.

Li, J., Jin, Y., & Zhou, P. (2025). Do banks price firms' biodiversity risk? Evidence from the Kunming declaration. *International Review of Financial Analysis, 2025*, 104557.

MSCI. (2023). *MSCI ESG ratings methodology: Biodiversity and land use key issue*.

Mulder, I., & Koellner, T. (2011). Hardwiring green: How banks account for biodiversity risks and opportunities. *Journal of Sustainable Finance and Investment, 1*(2), 103–120.

NGFS-INSPIRE. (2021a). *Biodiversity and financial stability: Building the case for action*. Network for Greening the Financial System and International Network for Sustainable Financial Policy Insights, Research and Exchange. https://www.ngfs.net/en/biodiversity-and-financial-stability-building-case-action

NGFS-INSPIRE. (2021b). *Biodiversity and financial stability: Exploring the action case* (NGFS occasional paper). Network for Greening the Financial System and International Network for Sustainable Financial Policy Insights,

Research and Exchange. https://www.ngfs.net/en/publications-and-statistics/publications/biodiversity-and-financial-stability-exploring-case-action
Posth, J. A., Schwendner, P., Laube, P., & Orpiszewski, T. (2024). *Bio-value-at-risk: A concept to assessing the implications of biodiversity risks on portfolio management using geospatial analysis*. Available at SSRN 4784271.
Roeder, R., & Utz, S. (2025). From diversity to confusion? The challenge of biodiversity footprint quantification. *Business Strategy and the Environment, 2025*, 1.
Schoenberger, E. (2002). The firm in the region and the region in the firm. In *The new industrial geography* (pp. 205–224). Routledge.
TNFD. (2023). Recommendations of the taskforce on nature-related financial disclosures. *Taskforce on nature-related financial disclosures.*
Tran, N. H., Fu, L., & Boehe, D. M. (2023). How does urban air pollution affect corporate environmental performance? *Journal of Cleaner Production, 383*, Article 135443.
WWF. (2024). *WWF biodiversity risk filter methodology documentation version 2.0, October 2024*. https://doi.org/10.5281/zenodo.13768225
Xin, W., Grant, L., Groom, B., & Zhang, C. (2025). Noisy biodiversity: The impact of ESG biodiversity ratings on asset prices. *Ecological Economics, 236*, Article 108662.

CHAPTER 5

The Role of Banks in Halting Biodiversity Loss

Abstract This chapter explores the changing role of banks in addressing biodiversity loss, shifting from a focus on risk management to a nature-positive finance approach. It begins by highlighting the ongoing financial need for biodiversity, which remains unmet, as well as the dominance of public funding and the negative impact of harmful subsidies. We examine a dual strategy to address this financial gap, which encompasses both disinvestment and engagement strategies. These strategies involve exclusion and divestment policies as well as active stewardship. Additionally, we discuss new channels and instruments for financing biodiversity, such as biodiversity-linked credit products, green, blue, or sustainability-linked bonds, conservation bonds, biodiversity credits, private equity, venture capital, debt-for-nature swaps, payments for ecosystem services, and blended or catalytic finance structures. The chapter concludes by identifying key obstacles to mobilising private capital at scale. It emphasises the urgent need for stronger collaboration between public and private sectors to align banking practices with the Kunming–Montreal Global Biodiversity Framework, ultimately aiming to achieve measurable nature-positive outcomes.

Keywords Conservation finance · Biodiversity finance · Nature-positive finance · Green finance · Blended finance

G. Rimo and S. Cosma, *Biodiversity and Banks*, Palgrave Studies in Impact Finance, https://doi.org/10.1007/978-3-032-17631-8_5

5.1 From Risk Management to Nature-Positive Finance

Earlier chapters have discussed how nature degradation and biodiversity loss can affect banks through multiple transmission channels, ultimately threatening financial stability. However, recognising biodiversity as a risk driver represents only one side of the equation. The other, increasingly relevant, dimension lies in the transformative potential of the banking sector to mobilise capital towards nature-positive outcomes. In other words, banks can (or should) move from merely ***managing exposure*** to *driving solutions*.

5.1.1 The Biodiversity Finance Gap

The urgency to address this issue arises from the evidence of a significant "biodiversity finance gap": the shortfall between existing financial resources and those needed to protect and restore nature effectively.

Despite the accelerating rate of biodiversity loss and growing awareness of the economic consequences of biodiversity loss, the allocation of capital to biodiversity conservation remains dramatically insufficient. In 2019, global financial flows directed towards biodiversity protection were estimated at only US$124–143 billion per year, compared with the US$722–967 billion that would be required annually to halt the ongoing loss of biodiversity by 2030. Collectively, these estimates indicate a biodiversity finance gap ranging from US$598 to US$824 billion annually (Deutz et al., 2020). The problem becomes more significant when we analyse the sources of funding for biodiversity conservation. As highlighted by Cosma et al. (2023), funding for biodiversity protection has long been provided by governments on a residual basis, only after they have addressed social needs and political challenges. Although a variety of financial instruments and mechanisms could be mobilised, the primary sources of funding for biodiversity conservation remain primarily dependent on grants, donations, and government budget allocations. Typical contributors include non-governmental organisations (NGOs), private corporations, and philanthropic foundations.

As shown in Fig. 5.1, the vast majority of biodiversity-related funding still originates from public sources. In 2019, domestic budgets and tax policy accounted for approximately 57% of total biodiversity finance (US$75–78 billion), while natural infrastructure investments (investments in ecosystems that perform infrastructure functions such as mangroves, protective forests, wetlands, and natural basins) represented about 20% (US$27 billion). The remaining share was distributed among smaller channels such as biodiversity offsets (projects financed by companies to offset the environmental impact of their activities), official development assistance (development aid for conservation: funds from wealthy governments to developing countries, e.g., through OECD-DAC), sustainable supply chains (corporate investments to make supply chains sustainable, certifications, traceability, and impact reduction), and green financial products (green bonds, sustainability-linked loans, nature-oriented ESG funds), each contributing less than 10% of the total. Private sector and market-based initiatives, including philanthropy, nature-based solutions, and carbon markets, remain marginal, jointly representing only a few percentage points.

This composition highlights the persistent reliance of global biodiversity financing on government expenditure, rather than on scalable private or blended finance mechanisms. In this context, the private financial sector, and particularly banks, appears as the *missing link* in the global effort to close the biodiversity finance gap.

At the same time, the figures on the biodiversity finance gap become even more alarming when compared with the vast capital flows that continue to be directed each year towards activities that harm nature. According to estimates from the OECD, government support that is environmentally harmful, including subsidies, exceeds USD 800 billion annually. These measures distort market signals and resource allocation, ultimately altering the structure of production and consumption within the economy (Matthews & Karousakis, 2022). Even if financial flows for biodiversity conservation are maximised, the global biodiversity finance gap will remain unresolved unless substantial efforts are made to accelerate the reform of biodiversity-harmful subsidies and to strengthen investment risk management practices within the financial sector (Deutz et al., 2020).

The Kunming–Montreal Global Biodiversity Framework explicitly sets the objective of progressively narrowing the estimated US$700 billion annual biodiversity finance gap (Willis, 2026). In order to reduce this gap, a twofold strategy emerges:

- **Reduce the need for conservation spending**: Limiting financial flows towards activities that harm nature lessens the amount of

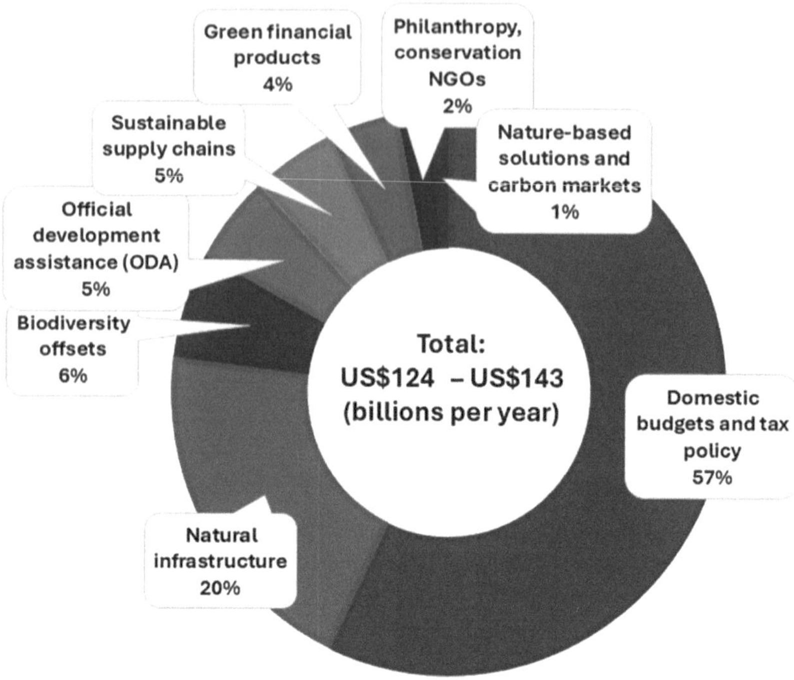

Fig. 5.1 Amount of funding for global biodiversity conservation in 2019 and analysis of financial flows (in billions of US dollars per year in 2019). *Source* Authors' elaboration on data from Deutz et al. (2020)

money needed to repair ecosystems. Reforming harmful subsidies and adopting more effective public and private practices can help prevent biodiversity loss at its source.

- **Increase funding for conservation**: Directing more private capital towards biodiversity-friendly investments. Fiscal incentives and targeted policies could significantly boost the amount of money flowing into conservation.

The EU Biodiversity Strategy for 2030, the Kunming–Montreal Global Biodiversity Framework (GBF), and the NGFS Conceptual Framework for Nature-Related Financial Risks all converge on the idea that halting and reversing biodiversity loss requires aligning financial systems with nature-positive objectives.

The concept of nature-positive finance, increasingly adopted by the TNFD (2023), UNEP FI, and the OECD, refers to financial strategies that intentionally direct capital flows towards activities that protect, restore, or sustainably manage biodiversity. It entails three progressive levels of action:

- **Avoid harming nature:** Avoid financing projects that significantly impact critical ecosystems; introduce environmental screening in due diligence; and recognise biodiversity as a potential source of credit, reputational, and legal risk. In essence, prevent financial flows from worsening nature loss.
- **Reduce the risks posed by nature:** Incorporate both quantitative and qualitative biodiversity metrics into lending and investment processes; prioritise counterparties that have credible impact management plans, and utilise specific metrics when assessing risk pricing; revise covenants and contractual structures to minimise exposure to high-impact sectors. The objective is to mitigate financial vulnerability resulting from degraded ecosystems or stricter regulations.
- **Contribute to nature restoration:** utilise innovative financial instruments; establish public–private partnerships for conservation; and invest in projects that enhance the health and functionality of ecosystems (e.g., reforestation, wetland restoration, and sustainable landscape management).

Banks therefore stand at the intersection of two imperatives: safeguarding their portfolios against nature-related risks and channelling capital towards the transition to a nature-positive economy. Bridging this divide requires innovation in financial instruments, credit policies, and impact measurement frameworks.

5.2 Redirecting Financial Flows: Disinvestment and Engagement Strategies

As just outlined, addressing the biodiversity finance gap requires not only the creation of new financing mechanisms but also the redirection of existing financial flows away from activities that damage nature. For banks and financial institutions, this means that nature-related strategies must operate on two complementary fronts: disinvestment, aimed at reducing exposure to biodiversity-negative assets, and engagement, designed to influence corporate behaviour and accelerate the transition of clients and portfolio companies towards biodiversity-friendly practices.

Indeed, banks influence the real economy through their role as financial intermediaries. By deciding what to finance and under which conditions, they shape the production structure and sustainability trajectory of entire sectors (Azizi, 2025). This influence can be exercised both negatively, through exclusion and divestment from harmful activities, and positively, through engagement and stewardship (Krueger et al., 2020). Together, these actions represent a form of ***indirect biodiversity governance***, in which financial actors leverage their capital allocation power to align markets with nature-positive objectives.

5.2.1 Exclusion and Divestment Policies

The first and most immediate approach available to banks is the adoption of sectoral exclusion and disinvestment policies. These involve limiting or eliminating financial relationships with companies or projects whose activities are demonstrably harmful to biodiversity, such as deforestation, illegal logging, unsustainable fishing, mining in protected areas, or land conversion of high ecological value.

An *exclusion approach* involves reducing the investable universe by explicitly excluding specific issuers, sectors, products, or countries. However, exclusions only influence future capital allocations. To effectively mitigate the negative impacts of existing investments, it is essential to establish a divestment policy that is consistent with the exclusion criteria. Such a policy addresses past investment decisions and the current composition of the portfolio. For example, following the COP 21, also known as the Paris Agreement, European banks have begun to divest from polluting companies (Reghezza et al., 2022). This is the result of policies that increase attention to the consequences of climate change, as well as the pressure of climate activism on reducing the supply of financial capital to the brown sectors (Cojoianu et al., 2021).

Several central international banks have begun translating their biodiversity commitments into concrete exclusion and risk management policies. BNP Paribas, for example, has introduced one of the most restrictive policies in the European banking sector to combat deforestation in the Amazon and Cerrado regions. Since 2019, the bank has prohibited the financing of clients producing or purchasing beef or soy from land deforested after specific cut-off dates, requiring full traceability of supply chains by 2025 and compliance with zero-deforestation commitments.[1]

Similarly, ING integrates biodiversity considerations into its Environmental and Social Risk (ESR) framework, applying enhanced due diligence to sectors with high potential for adverse impacts on nature, such as mining, forestry, and agriculture. The bank also develops sector-specific sustainability policies tailored to these sectors. For instance, ING explicitly states that it "does not finance projects or clients active in various high-value natural areas, including Ramsar Wetland areas, UNESCO World Heritage Sites, or areas designated for protection by the International Union for the Conservation of Nature (IUCN)".[2]

Together, these examples illustrate how large banks are starting to operationalise biodiversity risk management by restricting finance for ecosystem-damaging activities, tightening due diligence, and conditioning lending on verifiable nature-related performance standards.

Although exclusion policies may appear reactive, they serve as powerful signals that reshape market expectations. They help internalise externalities by increasing the cost of capital for biodiversity-negative sectors while redirecting resources towards more sustainable activities. Moreover, these policies reduce banks' own exposure to transition risks linked to biodiversity-harmful lending.

At the same time, exclusionary approaches have limits, and there is a debate about whether divestment or engagement is more effective (Cosma et al., 2024). Excessive reliance on divestment may reduce the ability of financial institutions to influence corporate behaviour, particularly in jurisdictions where access to alternative financing remains

[1] https://group.bnpparibas/en/press-release/bnp-paribas-defines-restrictive-policy-fight-deforestation-amazon-cerrado-regions.

[2] https://www.ing.com/Sustainability/Our-Stance/Nature-and-biodiversity.htm.

available. Moreover, divestment strategies may introduce greater risks, including reduced portfolio diversification and higher compliance costs (Cosma et al., 2024). For this reason, exclusion is increasingly complemented by engagement strategies, which represent a more proactive and collaborative avenue for achieving biodiversity outcomes.

5.2.2 *Engagement and Stewardship Strategies*

Engagement refers to the systematic process through which banks and investors use their financial leverage to influence the environmental and social practices of their clients and investee companies. Within the biodiversity context, engagement aims to enhance corporate management of nature-related risks, promote transparency regarding dependencies and impacts, and promote the adoption of science-based targets for biodiversity. Rather than withdrawing capital immediately, engagement aims to accompany the transition of high-impact sectors towards biodiversity-positive models.

Collaborative engagement initiatives have emerged as a response to the need for scale, consistency, and leverage in influencing corporate behaviour on biodiversity. By joining forces, financial institutions amplify their influence, reduce duplication of efforts, and create shared expectations for corporate transparency and action. In recent years, numerous global banks have undertaken joint initiatives that encourage the management and integration of risks arising from impacts on and dependencies on nature (Mundaca & Heintze, 2024).

The most prominent platform in this regard is the *Finance for Biodiversity Pledge* (FfB), launched in 2020 by 26 financial institutions and now comprising more than 200 institutions as of May 2025.[3] This growing community represents a network of global finance leaders committed to taking action on biodiversity. The Pledge commits signatories to five concrete actions:

- Collaboration and knowledge sharing among financial actors on biodiversity topics;
- Engagement with companies to encourage nature-positive business practices;

[3] https://www.financeforbiodiversity.org/ffb-pledge/.

- Assessment of biodiversity impact and dependencies;
- Setting biodiversity targets aligned with global frameworks like the GBF and the TNFD;
- Public reporting on progress every 2 years.

Under the Pledge, member institutions collaborate through thematic working groups. For example, the *Engagement with Companies* working group in 2022 released a guide to support investor engagement with companies on biodiversity, translating the goals and targets of the Global Biodiversity Framework (GBF) into specific engagement questions. In addition, a dedicated subgroup prepared detailed sectoral briefs covering 10 industries, which are accessible exclusively to members.[4]

Alongside the FfB, the *Nature Action 100* initiative focuses explicitly on coordinated engagement with the world's 100 most biodiversity-impactful companies.[5] Bringing together over 230 institutional investors and representing more than US$30 trillion in assets under management or in advisory roles, it targets 100 companies across sectors most critical to nature preservation—including biotechnology and pharmaceuticals, chemicals, retail, agribusiness, forestry, packaging, personal and household products, and metals and mining.

The initiative provides a structured framework for investors to coordinate engagement efforts, set clear expectations for corporate disclosure and action, and track progress through standardised benchmarks. By amplifying the investor voice and focusing on systemically important sectors, Nature Action 100 exemplifies how collaborative engagement can drive companies to integrate biodiversity considerations into strategic decision-making and risk management (ItaSIF, 2024).

Alongside these initiatives, other collaborative structures at the sectoral level are equally important for strengthening the management of risks related to nature and biodiversity. The *Equator Principles* (first launched in 2003 and grounded in the International Finance Corporation's environmental and social standards) constitute one of the earliest and most influential voluntary benchmarks for assessing, mitigating, and monitoring environmental and social risks in project finance. Adopted by more

[4] https://www.financeforbiodiversity.org/working-groups/engagement-with-companies/.

[5] https://www.natureaction100.org/.

than 120 financial institutions globally, they promote a harmonised due diligence approach across extensive infrastructure and resource-intensive projects.

Complementing this, the *Biodiversity Finance Initiative* (BIOFIN), established in 2012 by the United Nations Development Programme, aims to address the global "biodiversity finance gap"—estimated at USD 598–824 billion per year (Deutz et al., 2020). BIOFIN encourages banks to assess their balance sheet exposure to ecosystem-service-related risks and redirect capital flows towards activities and sectors that contribute to biodiversity conservation and restoration.

More recently, the *Principles for Responsible Banking* (PRB), launched in 2019 under the UNEP Finance Initiative, provides a strategic framework for aligning banking practices with the Paris Agreement and the Sustainable Development Goals. Notably, Principle 2 calls for banks to identify and mitigate the negative impacts associated with their operations, products, and services, including those related to nature and biodiversity. Today, more than 130 banks have committed to the PRB, signalling broad industry engagement around nature-related accountability and action (Mundaca & Heintze, 2024).

Individual engagement occurs when a financial institution engages directly and bilaterally with its clients or investee companies to encourage the adoption of biodiversity-friendly practices. Rabobank's approach to regenerative agriculture serves as a concrete engagement strategy for promoting biodiversity within the agri-food value chain. Instead of relying on exclusion or divestment, the bank actively collaborates with wholesale clients and farmers to encourage practices that restore soil health, enhance ecosystem resilience, and protect biodiversity. This strategy includes financial incentives, such as preferential lending terms for regenerative projects, and technical engagement through tools like the Open Soil Index and Biodiversity Monitor. Rabobank also forms collaborative partnerships with food companies, such as McCain Foods, to integrate biodiversity objectives into their supply chain practices. By aligning financial, technical, and operational measures, Rabobank drives systemic change in sectors crucial for halting the loss of nature.[6]

AXA Investment Managers' biodiversity engagement strategy integrates active stewardship into its listed equity impact approach, aiming to

[6] https://www.rabobank.com/about-us/impact/article/011455554/how-wholesale-companies-can-help-scale-up-the-regenerative-movement.

influence corporate behaviour towards nature-positive outcomes. Beyond selecting companies that already contribute to biodiversity preservation (*asset contribution*), AXA IM systematically engages with at least half of its portfolio holdings to encourage the adoption of measurable biodiversity targets, enhanced disclosure practices, and the reduction of negative environmental externalities (*investor contribution*). Engagement activities include dialogue, proxy voting, and performance monitoring aligned with frameworks such as the Taskforce on Nature-related Financial Disclosures (TNFD) and the Kunming–Montreal Global Biodiversity Framework.[7] By combining capital allocation with direct corporate engagement, AXA IM seeks to drive tangible improvements in how firms assess, manage, and report their impacts and dependencies on nature.

The coexistence of individual and collective engagement approaches allows for a multi-layered approach: collaborative initiatives, such as the FfB Pledge and Nature Action 100, address systemic barriers and promote convergence of expectations across the market, while bilateral engagement fosters customised change at the firm level. When combined, they form a vital link between proactive biodiversity finance and risk management, transforming the influence of financial institutions into a driving force for genuine ecological change.

5.3 Financing Biodiversity: Channels and Instruments

If disinvestment and engagement represent the first step towards redirecting capital away from biodiversity-harming activities, the complementary challenge lies in mobilising new capital towards biodiversity-positive outcomes. According to Elliott and Löfgren (2022), the banking sector still lacks a critical reflection on the indirect negative impacts stemming from its exposure to *brown companies*, which are environmentally harmful companies that continue to attract significant financing and investment from banks and other financial institutions. The global financial system still allocates only a small fraction of resources to nature protection, despite growing evidence that healthy ecosystems are a foundation for economic stability and long-term value creation. For banks, this implies

[7] https://www.axa-im.com/sustainability/insights/why-and-how-integrating-biodiversity-listed-equity-strategies.

not only integrating biodiversity into risk frameworks but also developing financial instruments that actively support conservation, restoration, and sustainable use of natural assets.

Recently, academics have started talking about *Conservation* and *Biodiversity finance*, referring to an emerging domain that mobilises private capital in two main ways:

1. By addressing risks and dependencies arising from adverse changes in nature, including implications for creditworthiness assessment;
2. By structuring investments that channel financial resources towards biodiversity restoration projects and activities with measurable conservation outcomes (Fenichel & Dean, 2024; Cosma et al., 2023; Flammer et al., 2025).

Beverdam et al. (2025) define biodiversity restoration projects as "*projects explicitly addressing the impact of human activities on ecosystems, by conserving, restoring or rehabilitating the functioning of ecosystems*", and propose a classification of several financial instruments from the private sector that can help channel private funds to projects that may advance or threaten biodiversity. Building on their taxonomy, the instruments identifiable as private or mixed can be reorganised into three functional channels reflecting the operational architecture of biodiversity finance (Table 5.1). The first channel, ***biodiversity-linked credit instruments***, encompasses loans and guarantees that incorporate nature-related performance metrics into pricing or covenant structures. The second channel, ***bonds and capital-market products***, groups market-based securities and equity vehicles that mobilise investment for biodiversity outcomes. Finally, ***blended and catalytic finance structures*** include arrangements that combine concessional or public funds with private capital to mitigate risks, enhance transparency, and attract additional investors.

This functional classification synthesises the dispersed set of instruments presented by Beverdam et al. (2025). It aligns with the emerging triad through which financial institutions are increasingly linking financial performance with biodiversity impact: credit, capital-market, and blended mechanisms.

Table 5.1 Classification of private biodiversity finance instruments according to three functional channels

Category	*Economic–financial logic*	*Examples*
(A) Biodiversity-linked credit instruments	Credit and loan products that embed biodiversity or nature-related performance indicators into their pricing structure, covenants, or repayment conditions. These mechanisms align lending practices with measurable environmental outcomes	Blue/green loans Sustainability-linked loans Syndicated loans
(B) Bonds and capital-market products	Market-based and investable securities that channel resources towards biodiversity conservation or restoration. Their cash-flow structure, coupon, or underlying asset is tied to ecological performance or nature-positive activities	Green/blue bonds Sustainability-linked bonds Conservation bonds Equity and venture capital Biodiversity credits
(C) Blended and catalytic finance structures	Financial architectures combining public, philanthropic, and private capital to de-risk and crowd-in private investment. They include mechanisms that reduce informational and return asymmetries, enabling the scalability of biodiversity projects	Landscape and collective investment funds Debt-for-nature swaps Payment for ecosystem services (PES)

Note The table provides a classification of the main private or mixed financial instruments useful for channelling capital towards biodiversity conservation projects. *Source*: Authors' elaboration on a first taxonomy provided by Beverdam et al. (2025)

5.3.1 *Biodiversity-Linked Credit Instruments*

The banking sector's primary leverage point lies in its role as a credit intermediary. By embedding biodiversity or nature-related indicators into loan pricing, covenants, and credit assessments, banks can directly influence corporate behaviour and incentivise measurable ecological performance.

Green/Blue Loans

Green loans are financial instruments whose *proceeds* are earmarked to support projects and initiatives that generate positive environmental outcomes, foster climate change mitigation, and advance sustainable development (Di Tommaso et al., 2025). In a typical green loan arrangement, a private borrower obtains credit from one or more banks in exchange for specific commitments, beyond the standard contractual clauses, to allocate the proceeds exclusively to projects and assets that generate measurable environmental benefits, including climate change mitigation and biodiversity enhancement. The funds are managed through revenue management and monitoring mechanisms, and the client provides the bank with an annual report on their allocation and environmental impacts, including avoided CO_2 emissions.

These elements, including the use of proceeds, project evaluation, management of proceeds, and reporting, enable the loan to be classified as green, ensuring that the financing generates verifiable environmental benefits consistent with international best practices. From an economic perspective, pricing on green loans may be more favourable for borrowers compared to conventional loans, reflecting the assumption that green assets are more resource-efficient and therefore present a lower default risk. For instance, a green building may entail lower heating and cooling costs than a traditional one, thereby reducing the borrower's credit risk and allowing lenders to offer better financing terms. For example, the results of a large survey conducted by the European Banking Authority (EBA) show that over 70% of European banks apply lower rates for green loans than for traditional loans, and that "credit institutions typically offer up to two percentage points discount on the interest rate for loans meeting determined sustainability criteria" (European Banking Authority, 2023).

The development of the green loan market has been facilitated by the Green Loan Principles (GLP), established jointly by the Loan Market Association (LMA), the Loan Syndications and Trading Association (LSTA), and the Asia Pacific Loan Market Association (APLMA). Although their adoption is voluntary, the principles recommend external verification to ensure credibility and transparency.

A more specific variant is the *blue loan*, explicitly targeting the protection and sustainable use of marine and freshwater ecosystems. DBS Bank has pioneered the use of blue loans in Asia, exemplified by its financing agreement with Far Eastern New Century Corporation (FENC) in Taiwan.[8] The loan, amounting to approximately US$33 million, was designed to fund projects that contribute to the protection and restoration of marine ecosystems, including ocean plastic recycling, wastewater management, and the development of ocean-friendly products. The initiative illustrates how financial institutions can extend the green loan concept to the blue economy, linking credit provision to measurable outcomes for marine conservation and water sustainability.

Sustainability-Linked Loans

While green and blue loans belong to the *use-of-proceeds* family of sustainable credit products, Sustainability-linked loans (SLLs) differ in that they are not tied to a specific use of funds, but rather to the *borrower's overall performance* against predefined sustainability or biodiversity key performance indicators (KPIs). These loan agreements contain covenants that reward borrowers with lower interest rates when they meet predefined sustainability targets, or penalise them with higher rates if those goals are not achieved (Du et al., 2023). SLLs thus introduce a performance-based incentive mechanism that internalises nature-related risks and opportunities within credit relationships. Like the green loans, the SLL market also operates under some voluntary guidelines set by the *Sustainability-Linked Loan Principles* (SLLP).

A notable example is the €750 million 5-year SLL granted in 2020 by BNP Paribas to the Finnish pulp and paper company UPM. The interest rate of the loan is linked to UPM's performance against two environmental indicators: (i) achieving a *net-positive impact on biodiversity* in the forests it manages in Finland, and (ii) reducing CO_2 emissions from fuels and electricity by 65% by 2030. Rabobank's *Planet Impact Loan* represents another example of SLL with biodiversity-related performance targets.[9] Rather than restricting the use of proceeds to specific green projects, the loan ties its pricing to the borrower's environmental performance as measured by the *Biodiversity Monitor*, a set of science-based KPIs assessing land use, soil quality, and greenhouse gas emissions. Farmers achieving higher biodiversity scores benefit from lower interest rates, creating a direct financial incentive for conservation.

[8] https://www.dbs.com.sg/corporate/insights/blue-finance-for-sustainable-blue-economy.

[9] https://www.accountingforsustainability.org/en/knowledge-hub/case-studies/rabobank--building-biodiversity-with-impact-loans.html.

Empirical evidence shows that the growing prevalence of sustainability-linked clauses is also evident in **syndicated loans**. ING reports that a majority of syndicated financings for their large corporate clients already contain a sustainable element.[10] Syndicated loans, i.e., financing provided by a consortium of lenders, offer an effective vehicle for biodiversity-linked lending in large-scale corporate or infrastructure transactions. The syndicate leader plays a crucial role in developing tailor-made sustainability documentation and KPIs, coordinating among banks, and bridging multiple specialist units (loan capital markets, sustainable finance, and sector specialists). In sectors where biodiversity impacts are material, such as agriculture, extractives, or infrastructure, syndicated loans can facilitate the alignment of multi-lender financing with nature-positive objectives, while delivering diversified risk allocation among lenders.

Bonds and Capital-Market Products

Capital-market products play a pivotal role in scaling biodiversity finance by mobilising large pools of institutional capital. Bonds and related instruments enable issuers to mobilise large pools of capital from institutional investors, while incorporating biodiversity outcomes into the structure of the financial product itself. These mechanisms share a common goal, i.e., to link capital-market returns with ecological performance, but differ in terms of *use of proceeds*, *performance logic*, and *risk-transfer design*.

Green/Blue Bonds

Following Flammer (2021), green bonds can be defined as "bonds whose proceeds are committed to finance environmental and climate-friendly projects, such as renewable energy, green buildings, or resource conservation". These debt instruments, issued by public or private entities, may be offered publicly (listed on exchanges and available to a broad base of investors) or privately placed with a limited number of institutional participants.

Research indicates that green bonds typically exhibit a slight premium over similar conventional bonds, a phenomenon commonly referred to as the *greenium*. This green premium reflects investors' willingness to accept slightly lower yields in exchange for environmental benefits and reputational value. Following this intuition, green bonds may represent

10 https://www.ingwb.com/en/insights/sustainability/orange-blog/sustainable-syndicated-loans-need-for-expertise.

a cheaper source of financing (Flammer, 2021). A report by Climate Bonds Initiative (2023), for example, found that all types of environmentally labelled bonds presented lower average spreads relative to their vanilla comparables. However, this is not confirmed by empirical research, finding no evidence that green bonds are priced at a premium compared to nongreen bonds (Flammer, 2021; Larcker & Watts, 2020).

To ensure transparency and consistency across issuers, the financial industry developed the Green Bond Principles (GBP) under the auspices of the International Capital Market Association (ICMA). These voluntary principles aim to ensure greater standardisation in the market. On the one hand, initiatives like the Climate Bonds Initiative (CBI) have established sector-specific taxonomies and certification criteria to help issuers and investors evaluate whether a project qualifies as "climate-friendly". On the other hand, no dedicated criteria currently exist for biodiversity-related investments. However, the ICMA GBP explicitly lists "terrestrial and aquatic biodiversity conservation" among the eligible green project categories to which green bond proceeds may be allocated (ICMA, 2023). In 2021, the Italian electricity transmission operator Terna S.p.A. issued a green bond to finance projects aimed at reducing land consumption and mitigating terrestrial biodiversity loss. The bond proceeds were allocated to the demolition of existing overhead power lines and pylons, particularly in environmentally sensitive areas, including national parks, wetlands, and protected landscapes.[11]

The proportion of green bonds allocating proceeds to biodiversity conservation has increased markedly in recent years, rising from 5% of issuances in 2020 to 16% in 2023. However, the overall pace remains modest.[12] At the same time, a new wave of "blue" debt instruments is emerging, driven by the same structural logic as green bonds. **Blue bonds** are debt securities specifically designed to raise funds for marine- and water-based projects with positive environmental, economic, and climate impacts.[13]

For example, the government of the Seychelles issued the world's first sovereign blue bond in 2018, raising US$15 million to finance

[11] Information come from the GREEN BOND REPORT 2021 of Terna S.p.A.

[12] https://www.im.natixis.com/en-latam/insights/fixed-income/2024/green-bonds-to-preserve-biodiversity.

[13] https://www.worldbank.org/en/news/feature/2018/10/29/sovereign-blue-bond-issuance-frequently-asked-questions.

marine conservation and sustainable fisheries in its 1.4 million km^2 marine territory.

These developments suggest that the green bond model, initially designed for climate mitigation, is progressively evolving into a broader tool for financing biodiversity-positive activities. By offering investors a low-risk, standardised, and transparent instrument, green bonds can mobilise capital at scale for conservation and ecosystem restoration. However, the absence of harmonised metrics and certification standards for biodiversity projects continues to limit their full potential, underscoring the need for a more integrated taxonomy that reflects the diversity and complexity of nature-related finance.

Box 5.1—Illustrative Cases of Blue Bond Issuances: The Republic of Seychelles and the Nordic Investment Bank

Case A: Seychelles Sovereign Blue Bond (2018)

In October 2018, the Republic of Seychelles became the first country in the world to issue a sovereign blue bond, raising US$15 million through a 10-year issuance aimed at financing marine conservation and sustainable fisheries management.[14] The transaction was structured with the support of a US$5 million partial guarantee from the World Bank and a US$5 million concessional loan from the Global Environment Facility (GEF), which reduced the borrowing costs for the issuer.

Proceeds were allocated to:

- promoting the sustainable management of the country's Exclusive Economic Zone (EEZ), covering approximately 1.4 million km^2 of marine area;
- expanding marine protected areas (MPAs) and funding community-based fisheries initiatives within the national *Blue Economy Strategy*.

This pioneering operation marks a milestone in sovereign sustainable finance, demonstrating how small island developing states can effectively integrate biodiversity conservation into their debt management and fiscal policies.

[14] https://www.worldbank.org/en/news/press-release/2018/10/29/seychelles-launches-worlds-first-sovereign-blue-bond.

Case B: Nordic Investment Bank (NIB)—Nordic-Baltic Blue Bond (2019–2020)

In January 2019, the Nordic Investment Bank (NIB) issued its first Nordic-Baltic Blue Bond, raising SEK 2 billion ($\approx$ EUR 200 million) under its *Environmental Bond Framework*.[15] A second issuance followed in 2020 for SEK 1.5 billion, maturing in 2025 and oversubscribed by more than 2.5 times the offer volume, confirming strong investor demand.

The proceeds were earmarked exclusively for projects in the Nordic–Baltic region addressing:

- sustainable water management and wastewater treatment;
- marine-pollution prevention;
- adaptation to climate change and flood resilience in the Baltic Sea basin.

Through these issues, NIB demonstrated how multilateral development banks can utilise blue bonds to finance regional environmental infrastructure and promote transboundary cooperation for the restoration of marine ecosystems. The bonds comply with the ICMA Green Bond Principles and align with the emerging taxonomy of "blue" use-of-proceeds instruments, bridging environmental goals and capital-market mechanisms.

Sustainability-Linked Bonds

Sustainability-linked bonds (SLBs) have recently emerged as a flexible and performance-oriented instrument within the sustainable debt market. Unlike traditional green bonds, which require issuers to allocate proceeds exclusively to environmentally beneficial projects, SLBs are not constrained by the specific use of funds. Instead, they are characterised by a financial structure that ties the bond's coupon payments to the achievement of predefined sustainability performance targets (Berrada et al., 2022). These targets are typically expressed through measurable key performance indicators (KPIs), such as reductions in greenhouse gas emissions, improvements in energy efficiency, or enhancements in biodiversity management. Should the issuer fail to meet the agreed-upon targets within the specified timeframe, a coupon step-up clause is activated, resulting in higher interest payments to investors. This mechanism

[15] https://www.nib.int/news/nib-issues-first-nordic-baltic-blue-bond.

aligns the financial incentives of issuers with their sustainability commitments, fostering accountability and signalling credibility to the market. This structure shifts the focus from the *use of proceeds* to *issuer behaviour*.

The ICMA Sustainability-Linked Bond Principles (SLBP) provide the key market framework for these instruments, setting guidance on KPI selection, calibration of Sustainability Performance Targets (SPTs), reporting, and external verification. To ensure credibility, issuers must demonstrate that the chosen KPIs are material, measurable, and ambitious, and a third party independently verifies that performance.

Although still primarily dominated by climate-related targets such as emission reduction, renewable energy, and energy efficiency, the potential usefulness of SLBs for biodiversity conservation through the incorporation of biodiversity-related KPIs is increasingly recognised (Agliardi & Agliardi, 2025). Such KPIs may include goals like achieving zero deforestation in supply chains, restoring natural habitats, or expanding certified sustainable land areas.

Conservation Bonds

Following the definition provided by Beverdam et al. (2025), conservation bonds can be defined as follows: "Bonds (debt) issued to finance a return-generating project. Interest payments on the bonds finance nature conservation operations, rather than flowing to the investor. When a certain pre-specified goal is achieved, the investor receives a payment". Their distinguishing feature lies in the direct link between financial returns and measurable conservation outcomes, rather than in the use of proceeds (as in green bonds) or in issuer-level performance (as in SLBs). While green bonds channel capital towards environmentally beneficial projects and SLBs reward or penalise the issuer depending on the achievement of predefined sustainability targets, conservation bonds shift the focus to *nature itself* as the performance metric.

They are structured as *pay-for-success* or *outcome-based* instruments: investors provide upfront capital for conservation activities, and returns are contingent on the verified achievement of ecological outcomes, such as increases in wildlife populations, hectares of restored forest, or reductions in pollutant loads. In some cases, the success payment is financed not by the issuer but by a third party (e.g., a multilateral development bank, a philanthropic foundation, or a global environmental fund), which compensates investors once the conservation targets are met.

The Wildlife Conservation Bond (WCB), also known as the *Rhino Bond*, represents one of the most innovative financial mechanisms ever designed to link capital-market instruments with measurable biodiversity outcomes. Issued by the World Bank (International Bank for Reconstruction and Development, IBRD) in March 2022, the 5-year bond raised US$150 million from institutional investors to finance the conservation of black rhinoceros populations in South Africa. This conservation bond works as follows:

1. Investors purchase the bond from the World Bank, which acts as issuer and intermediary, channelling private capital towards biodiversity conservation projects in South Africa.
2. The Global Environment Facility (GEF)[16] allocates up to US$13.76 million in performance-based outcome payments that will reward investors if the conservation targets are met.
3. Independent parties verify changes in rhino populations across two protected areas: the Addo Elephant National Park and the Great Fish River Nature Reserve.
4. The South African National Parks Authority (SANParks) implements conservation measures on the ground, using the bond proceeds.

Credit Suisse played a pivotal role as sole structurer and joint bookrunner (together with Citibank), designing the financial architecture that made the instrument compatible with investor expectations and biodiversity-linked performance metrics. In particular, the bank contributed to defining the *pay-for-results* mechanism by which the bond's "success payments" are triggered.[17] Through its structuring and distribution activities, Credit Suisse acted as a financial intermediary translating ecological performance into investor-grade financial returns, demonstrating how investment banks can operationalise nature-positive finance within the capital-market ecosystem. Figure 5.2 summarises the transaction structure of the Rhino Bond.

[16] The Global Environment Facility (GEF) is an International partnership fund bringing together 186 member countries, the leading UN agencies, the World Bank, regional development banks, NGOs and civil society organisations.

[17] https://www.thebanker.com/content/a1b9a7c7-ff2a-5000-bb2f-de0d8e1e6e70.

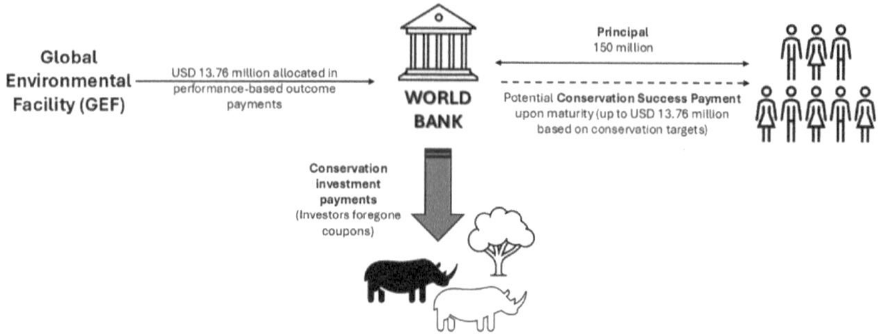

Fig. 5.2 Transaction structure of the rhino bond. *Source* Authors' elaboration on a case study by the World Bank[18]

Investors receive no periodic coupon payments. Investors' forgoing of coupons creates the economic space for conservation payments to be activated, allowing capital to be allocated towards rhino protection. Instead, at maturity, they are repaid the principal in full by the World Bank. They may receive an additional "Conservation Success Payment" financed by the GEF, depending on verified conservation results (World Bank, 2022). The success payment depends on the average annual net growth rate of the black rhino population in two South African protected areas (Table 5.2).[19]

Although the market remains nascent, these instruments illustrate how performance-based finance can complement green and sustainability-linked models, especially for projects that generate high ecological value but limited short-term revenue. As biodiversity metrics become more standardised and verification frameworks are established, conservation bonds are likely to evolve into a key pillar of the nature-positive finance ecosystem.

Private Equity and Venture Capital for Biodiversity
Private equity and venture capital funds are increasingly recognised as important mechanisms for mobilising private capital towards biodiversity

[18] https://thedocs.worldbank.org/en/doc/7039bd837e60e484fb3a93ea63951306-0340022022/original/CaseStudy-WildlifeConservationBond.pdf.

[19] https://hive.greenfinanceinstitute.com/gfihive/revenues-for-nature/case-studies/the-wildlife-conservation-bond-the-rhino-bond/.

Table 5.2 Investor payoff structure of the Wildlife Conservation Bond ("Rhino Bond")

Verified average annual growth of the black rhino population	*Conservation success payment (per USD 1000 invested)*	*Total amount disbursed by the global environment facility (GEF)*
$0\% < X \leq 2\%$	USD 36.69	≈ USD 5.5 million
$2\% < X \leq 4\%$	USD 73.38	≈ USD 10.9 million
$X > 4\%$	USD 91.73	≈ USD 13.76 million
$X \leq 0\%$	No payment	0

Note The table illustrates the relationship between conservation outcomes, specifically the growth of the black rhino population, and the financial returns on the bond. *Source* Authors' elaboration on data from International Bank for Reconstruction and Development (2022)

conservation and nature-positive innovation. Unlike debt-based instruments, which rely on fixed repayment structures, these equity vehicles channel long-term, patient capital into companies, natural assets, or projects that generate both financial returns and measurable ecological outcomes.

Private equity funds can contribute to biodiversity objectives by acquiring or financing firms that manage natural resources sustainably, such as those in forestry, regenerative agriculture, and water management, and by using their ownership power to influence corporate strategies and change business models (soil regeneration, deforestation-free agricultural supply chains, etc.). This instrument aims to enhance the environmental performance of existing companies and generate returns through cash flows (sustainable wood, carbon credits, and ecosystem services) in the medium to long term.

For instance, *The Nature Conservancy's NatureVest* platform has raised over US$130 million in combined equity and debt to finance the Cumberland Forest Project in the United States, protecting over 100,000 hectares of forestland while generating revenues from sustainable timber, carbon credits, and recreation.[20] Similarly, *Mirova Natural Capital* invests in large-scale restoration projects across Latin America and Africa,

[20] https://www.environmental-finance.com/content/awards/sustainable-investment-awards-2020/winners/impact-fund-of-the-year-the-nature-conservancys-sustainable-forestry-fund.html.

with a focus on landscape rehabilitation and sustainable commodity value chains.

Venture capital funds, by contrast, play a complementary role by financing early-stage innovations that enable biodiversity monitoring, restoration, and sustainable production. These include, for example, *Regeneration.VC*, which invests in circular materials and regenerative technologies, aims to finance technical interventions that address GHG emissions, material resources, or biodiversity within the corporate value chain of consumer industry businesses.[21] Through such initiatives, venture capital helps bridge the funding gap between conservation science and scalable business models.

Collectively, these instruments illustrate how equity-based finance can act as a catalyst for private sector engagement in biodiversity, aligning profit motives with conservation outcomes and accelerating the transition towards a nature-positive economy.

Biodiversity Credits

Emerging biodiversity finance instruments, such as biodiversity credits (also known as *biocredits*), are innovative financial tools designed to convert measurable conservation outcomes into tradable units that can attract private capital. A biodiversity credit represents a quantified gain or avoided loss of biodiversity, such as an improvement in species richness, habitat quality, or ecological connectivity, verified against a predefined baseline and expressed through standardised metrics. Unlike ***biodiversity offsets***, which compensate for ecological damage caused by specific activities, biodiversity credits operate in voluntary markets and aim to generate ***net-positive*** outcomes for nature rather than merely neutralise harm (Croci et al., 2025; Ducros & Steele, 2022; Wunder et al., 2025). They thus function as incentive-based instruments for channelling financial resources towards conservation and restoration, without the equivalence principle that underpins offsetting schemes.

Credits are issued by project developers, such as conservation NGOs, local communities, or private landowners, who undertake long-term commitments to protect or restore ecosystems (Arjaliès et al., 2025). The operational process involves:

[21] https://impactassets.org/ia50/fund.php?id=a01RQ00000OWfdJYAT.

1. Baseline assessment: identification of the ecological condition and metrics at project inception.
2. Measurement of uplift: quantification of biodiversity improvement through standardised indicators (e.g., number of species, vegetation cover, pollinator abundance).
3. Verification and certification: third-party auditing ensures additionality, permanence, and non-leakage.
4. Registration and issuance: credits are recorded in digital registries, often blockchain-based, to ensure traceability and prevent double-counting.

Each credit typically covers a duration of 10–30 years and can be traded in voluntary markets. Proceeds are distributed through *benefit-sharing mechanisms*: empirical evidence shows that 60–80% of revenues in existing schemes (e.g., Terrasos in Colombia, Wallacea Trust, ValueNature) are directed to Indigenous Peoples and Local Communities, while the remainder supports project operation, monitoring, and certification (Croci et al., 2025).

Companies, impact investors, or philanthropists acquire biodiversity credits primarily voluntarily, motivated by a combination of strategic, reputational, and regulatory drivers. In this sense, biodiversity credits operate as corporate instruments for positive environmental accountability, providing measurable evidence of nature-positive action even when direct operational restoration is not feasible.

A concrete example of how financial institutions are beginning to engage in biodiversity credit markets comes from Swedbank, one of the largest banks in the Nordic region.[22] In 2023, Swedbank purchased 91 biodiversity credits issued by the cooperative Orsa Besparingsskog, which manages around 75,000 hectares of high-value boreal forest in central Sweden. The credits were generated on a 13-hectare site under a project designed in collaboration with the Swedish University of Agricultural Sciences (SLU) and verified through Earthly, which is a platform for measuring nature-based impacts.[23]

The project quantifies biodiversity improvements using indicators such as the abundance of rare species, the amount of deadwood (crucial for

[22] https://www.swedbank.com/sustainability/environment/biodiversity.html.

[23] https://earthly.org/blog/swedebank-purchase-bidiversity-credits.

many organisms), and the level of sunlight reaching the forest floor, each factor contributing to an overall biodiversity score. Credits are reassessed and reissued every 5 years to ensure permanence and transparency.

For Swedbank, the initiative forms part of its broader "finance for nature" strategy. The purchase provides an additional revenue stream for forest owners to preserve, rather than harvest, mature forest stands, while offering the bank a verified, reportable contribution to its *nature-positive* objectives.

This case illustrates how biodiversity credits can integrate ecological outcomes into mainstream financial decision-making, linking local conservation action with institutional capital flows.

Moreover, financial institutions can play a pivotal role in scaling biodiversity credits markets through complementary functions, such as:

1. Financing and intermediation. Banks can provide upfront capital to project developers, structure biodiversity credit funds, and aggregate verified projects into investable portfolios or securitised products, thus bridging small-scale conservation efforts with institutional investors.
2. Product innovation and portfolio integration. Biodiversity credits can be embedded in sustainability-linked bonds, also known as green structured products, where returns depend on verified biodiversity outcomes. Incorporating biodiversity credits into financial products also strengthens banks' alignment with emerging *nature-positive finance* frameworks and enhances the quality of their TNFD and CSRD disclosures.

Ultimately, biodiversity credits offer a means for the financial sector to monetise and mainstream the value of nature, transforming conservation from a philanthropic expense into a measurable, tradable, and reportable asset class.

5.3.2 *Blended and Catalytic Finance Structures*

Blended finance combines public, private, and philanthropic financial mechanisms within a unified framework. By combining concessional public resources with market-based private capital, it aims to subsidise and de-risk private investment, thereby improving the overall risk–return profile of nature-related projects (Beverdam et al., 2025; Flammer et al., 2025; Kedward et al., 2023). In particular, blended finance aims to enhance the attractiveness of nature-related investments for private actors. By leveraging public or philanthropic capital to improve project bankability, blended finance enables initiatives that might otherwise remain financially unviable and mobilises larger financial flows towards biodiversity conservation than could be achieved through public funding alone (Beverdam et al., 2025; Deutz et al., 2020).

As highlighted in the previous sections, several financial instruments designed to channel capital towards biodiversity conservation already incorporate blended finance components in their structure.

For instance, the Seychelles Blue Bond combined concessional loans and guarantees[24] from the Global Environment Facility (GEF) and the World Bank to attract private investment in sustainable fisheries. Similarly, the Rhino Conservation Bond, issued by the World Bank in collaboration with Credit Suisse, relied on performance-based outcome payments from the GEF to mitigate risk and encourage investor participation. These examples demonstrate that blending is not a discrete category of instruments but rather a financial logic already embedded across multiple products discussed in this chapter.

However, beyond these hybrid structures, a distinct group of mechanisms has emerged in which the architecture of the instrument is purposefully designed around blended finance principles.

In these cases, the combination of concessional, philanthropic, and commercial resources is not an ancillary feature but the defining element that enables biodiversity investments to occur at scale.

This family of instruments is specifically intended to de-risk, leverage, and crowd-in private capital where traditional market-based instruments (such as bonds or equity funds) remain insufficient.

[24] Guarantees act as catalytic de-risking tools that make biodiversity investments more attractive. By offering partial credit coverage or first-loss protection, public or philanthropic actors reduce the perceived risk for commercial lenders, encouraging them to finance projects that would otherwise remain unfunded.

Landscape and Collective Investment Funds
Landscape and collective investment funds represent one of the most operationally mature and institutionally credible forms of blended finance for biodiversity. They are designed to address the intrinsic fragmentation of conservation projects, which are often too small, localised, or risky to attract institutional investors individually. By pooling multiple initiatives within a standard investment vehicle, these funds achieve both financial diversification and ecological coherence at the landscape or ecosystem level (Beverdam et al., 2025; Harrer et al., 2023).

Their financial structure is typically layered, combining concessional, public, or philanthropic capital with commercial investment tranches. The concessional component absorbs the first losses or accepts lower returns, improving the risk/return profile for private investors and encouraging their participation. It is a multi-income approach, more resilient than individual projects. Returns come from various sources (depending on the fund): the sale of certified and premium agricultural products, sustainable timber, carbon credits, biodiversity credits, and ecosystem services (e.g., water), as well as income from supported local businesses (including cooperatives and SMEs).

This model enables public resources to serve as a catalyst, attracting significantly larger volumes of private investment in sustainable land use, agriculture, and ecosystem restoration. Beyond financial blending, landscape funds integrate technical assistance, project preparation, and monitoring components. These activities ensure that the funded projects deliver measurable ecological and social outcomes, strengthening accountability and investor confidence. The qualifying elements that distinguish them from other donors are therefore in the object of the investment (entire landscapes and supply chains), in the source of return, which is not unique but diversified, in the dominant risk capital made up of public, private, and philanthropic, and finally, a non-localised but systemic impact on the territory and community.

A leading case is the *Land Degradation Neutrality Fund* (LDN Fund), launched by Mirova (a Natixis Investment Managers affiliate) and the United Nations Convention to Combat Desertification (UNCCD), which operates as a blended finance vehicle designed to mobilise both public and private capital towards sustainable land management and restoration in developing countries. With a total capitalisation of approximately USD

208 million and a lifespan of 15 years, the LDN Fund exemplifies a collective investment structure blending concessional capital from development finance institutions with private investors' equity. This structure enhances the risk-return profile for private investors and enables the financing of projects that conventional commercial funds typically avoid due to long payback horizons or environmental risk.

Its public investors include significant development and environmental institutions such as the Agence Française de Développement (AFD), the European Investment Bank (EIB), the UK Department for Environment, Food and Rural Affairs (DEFRA), the Governments of Canada and Luxembourg, and the Global Environment Facility (GEF). These actors provide concessional capital and risk-mitigation functions, ensuring alignment with the UN Sustainable Development Goals. The private investors consist of institutional and corporate actors, including BNP Paribas Cardif, Natixis Assurances, Allianz, Garance Mutuelle, Fondaction, Fondation de France, Concordia University Foundation, L'Oréal, and BRED Banque Populaire. They invest in commercial or impact-oriented capital, seeking stable returns and measurable environmental outcomes.

The LDN Fund's portfolio includes 13 projects across three continents (Africa, Asia, and Latin America), with an average investment horizon of 7–10 years. Examples include:

- Terrasos (Colombia): Development of Habitat Banks and biodiversity credit markets in compliance with national environmental compensation schemes.
- Komaza (Kenya): Micro-forestry and community woodlots integrating smallholder farmers into sustainable timber value chains.
- Aqre Group (Madagascar): Production of natural pharmaceutical ingredients (artemisinin, quinine) through sustainable cropping on degraded land.

The Fund applies a rigorous ESG and impact framework, consistent with the UNCCD principles, IFC Performance Standards, and alignment with the SDGs. Projects are required to demonstrate measurable benefits in:

- Environmental impact: CO_2 sequestration, soil restoration, biodiversity protection.
- Social impact: employment, gender equality (2X Challenge criteria), inclusion of smallholders.
- Economic resilience: long-term income generation through sustainable value chains.

Performance is monitored using key impact indicators, and each investment is accompanied by an Environmental and Social Action Plan (ESAP) that has been agreed upon with the investee company.

In summary, landscape and collective investment funds play a dual role in biodiversity finance. Financially, they transform scattered conservation projects into investable portfolios that meet the requirements of institutional investors. Ecologically, they promote integrated landscape management, recognising the links between biodiversity, land use, and local livelihoods. By combining financial innovation, partnership governance, and measurable impact assessment, they embody the systemic ambition of nature-positive finance, connecting local restoration efforts with global capital markets.

Debt-for-Nature Swaps

Debt-for-nature swaps (DNS) are financial arrangements in which a portion of a country's external debt is forgiven, converted, or refinanced in exchange for a commitment to invest in biodiversity conservation or climate-related objectives (Cosma et al., 2023). The concept emerged in the mid-1980s as a response to the dual crises of external debt and environmental degradation affecting developing countries. The mechanism allows part of a nation's debt to be redeemed at a discount and the equivalent value in local currency to be channelled into domestic conservation activities (Occhiolini, 1990).

As documented by Fuller (1989), the first generation of DNS involved non-governmental organisations purchasing sovereign debt on the secondary market at a discount, negotiating with debtor governments to convert it into local-currency bonds or trust funds, and using the proceeds to finance conservation projects. In Bolivia, for instance,

Conservation International acquired USD 650,000 of debt at roughly 10 cents on the dollar in 1987, leading to the legal protection of 3.7 million acres of tropical forest and the creation of a local conservation fund. Similar operations followed in Costa Rica, Ecuador, the Philippines, Madagascar, and Zambia, where converted debt supported the establishment of national parks, reforestation programmes, ranger training, and environmental education. These early swaps were bilateral or philanthropic transactions, mainly funded by NGOs (such as WWF, The Nature Conservancy, and Conservation International) and donor governments, without mobilising private investment (Fresnillo, 2023). In the early days of debt-for-nature swaps, the primary creditors were mostly international commercial banks that held severely distressed sovereign debt. These banks agreed to sell the debt at substantial discounts, not out of a desire to support environmental initiatives, but because secondary market prices already indicated a high likelihood of default. As a result, it was more advantageous for them to liquidate the debt immediately rather than hold onto non-performing assets.

Over time, the DNS model evolved into more sophisticated and market-linked forms that combine public de-risking mechanisms and private capital, aligning with the definition of blended finance. In recent years, debt conversion has taken the form of Blue Bonds, Green Bonds, or other sustainability-linked sovereign instruments, which replace a portion of existing external debt with new securities whose proceeds are earmarked for conservation and restoration projects (Simeth, 2025). Table 5.3 illustrates selected country cases where sovereign debt conversions have been structured through thematic bonds.

Through this structure, DNS combine debt relief, concessional capital, and market financing, creating investable vehicles that link fiscal sustainability with measurable environmental outcomes. Modern swaps also involve development finance institutions (DFIs) providing partial guarantees or first-loss protection, thereby reducing sovereign and political risk. This de-risking enables commercial banks and institutional investors to participate as underwriters, arrangers, or buyers of the restructured bonds, thereby aligning their financial and environmental objectives.

Table 5.3 Sovereign debt conversions using thematic bonds: selected debt-for-nature swap cases

Country	*Year*	*New instrument from conversion*	*Conservation Commitments*
Seychelles[25]	2018	Sovereign blue bond	Expansion of marine protected areas; sustainable fisheries management; the target of protecting 30% of the Exclusive Economic Zone
Belize[26]	2021	Blue loan/blue bond	Protection of at least 30% of Belize's marine areas; long-term financing through a dedicated Conservation Fund
Barbados[27]	2022	Blue bond (impact bond)	Marine conservation, coastal restoration, climate resilience, and support for the island's blue economy
Ecuador–Galápagos[28]	2023	Galápagos marine bond (blue/nature bond)	Protection of Galápagos marine areas; management of marine protected areas; financing of the Galápagos Life Fund
Bahamas[29]	2024	Nature bond (blue-focused)	Marine conservation, mangrove and seagrass restoration, climate resilience, and strengthening of marine protected areas

Note This table provides some notable examples of public debt restructuring that involved the simultaneous issuance of thematic bonds aimed at restoring nature. *Source*: Authors' elaboration

25 https://thecommonwealth.org/case-study/case-study-innovative-financing-debt-conservation-swap-seychelles-conservation-and.

26 https://www.debtfornature.org/case-studies-blog/belize.

27 https://www.nature.org/en-us/newsroom/tnc-announces-barbados-blue-bonds-debt-conversion/.

28 https://www.iadb.org/en/news/ecuador-completes-worlds-largest-debt-nature-conversion-idb-and-dfc-support.

29 https://www.nature.org/en-us/newsroom/tnc-announces-new-nature-bonds-project-bahamas/.

Box 5.2. The Belize 2021 Debt-for-Nature Swap: The Landmark "Blue Bonds for Ocean Conservation" Case **1. Background and Rationale**

Belize entered the COVID-19 crisis with one of the highest public-debt burdens in the Caribbean (debt-to-GDP ratio of 133% in 2020). A single Eurobond, known as the Superbond, represented almost all of the country's external commercial debt and traded at a deep discount due to concerns over Belize's fiscal sustainability. At the same time, Belize hosts critical marine ecosystems, including a large portion of the Mesoamerican Reef, whose conservation requires stable, long-term finance. This context made Belize an ideal candidate for an innovative Debt-for-Nature Swap capable of simultaneously reducing sovereign debt stress and securing conservation outcomes.

2. The DNS in a Nutshell

- Belize repurchases its USD 553 million "Superbond" at a deep discount (55 cents per dollar).
- It does not use its own resources to do so; instead, it receives a new "Blue Loan" from a TNC (The Nature Conservancy) owned special-purpose vehicle (BBIC).
- The Blue Loan is financed by private capital, raised through the issuance of Aa2-rated Blue bonds, purchased by institutional investors.
- The elevated rating is made possible by a political risk insurance guarantee from the U.S. Development Finance Corporation (DFC).
- Credit Suisse acts as arranger and lead distributor of the blue bonds, structuring the instrument and placing it with institutional investors. The bank converts the DFC-enhanced sovereign exposure into an investable, investment-grade ESG product, thereby enabling private capital to enter the transaction at scale.
- Belize achieves debt reduction and debt-service savings; in exchange, it commits to channel part of these savings to marine conservation for 20 years and adopts stringent conservation obligations.

It is therefore a second-generation Debt-for-Nature Swap, combining commercial debt refinancing, credit enhancement, parametric climate insurance, and milestone-linked penalties.

3. Belize's Environmental Commitments

The agreement includes several commitments for the Government of Belize:

- Protect 30% of Belize's ocean by 2026
- Complete a Marine Spatial Plan by 2026
- Protect mangroves within the Belize Barrier Reef Reserve System
- Revise the Integrated Coastal Zone Management Plan
- Submit three protected areas to the IUCN Green List

4. Results of the Swap

Table 5.4 synthesises the main financial and conservation benefits achieved through this innovative Debt-for-Nature Swap.

Payments for Ecosystem Services

Payment for Ecosystem services (PES) schemes represent one of the earliest and most widely adopted mechanisms linking financial incentives with the preservation of natural capital. Following the definition provided by Wunder (2005), "the core idea of PES is that external ecosystem services beneficiaries make direct, contractual and conditional payments to local landholders and users in return for adopting practices that secure ecosystem conservation and restoration" (Wunder, 2005). In other words, those who benefit from ecosystem services, such as clean water, carbon sequestration, pollination, or landscape beauty, should compensate those who maintain or enhance these services. In essence, PES translate the non-market value of nature into a measurable and tradable transaction.

Landowners, farmers, or local communities commit to specific conservation or restoration practices such as maintaining forest cover, preserving wetlands, or adopting agroecological methods, in exchange for a monetary payment or favourable financing terms. Payments may come from public authorities, private corporations, utilities, or financial intermediaries seeking to offset their environmental footprint or strengthen supply chain resilience. Wunder (2005) presents four main categories of ecosystem services underpinning the majority of PES schemes:

1. **Carbon sequestration and storage:** PES mechanisms focused on carbon involve payments to landowners or local communities for activities that enhance carbon capture or avoid emissions. *Example:* A Northern electricity company pays tropical farmers to plant and maintain additional trees, thereby compensating for its residual carbon emissions (i.e., those still present after internal reduction).

Table 5.4 Financial and conservation benefits deriving from the debt-for-nature swap

Category	*Benefits*	*Details*
Financial benefits for Belize	Debt stock reduction	USD 189 million reduction (from USD 553 m to USD 364 m)
	Debt-service savings	Approx. USD 200 million over 20 years
	Avoided reinstatement	USD 58.4 million "reinstated principal" avoided from prior restructuring
	Maturity extension	+6.5 years on average maturity of public debt
	Sovereign credit improvement	S&P upgrade from *Selective Default* to B−
	Improved debt sustainability	IMF acknowledges the swap as a key driver of improved debt sustainability indicators
	Lower default risk	Reduced short- and medium-term refinancing and liquidity risk
Conservation benefits	Long-term conservation financing	Approx. USD 180 million over 20 years
	Conservation endowment	National endowment created (USD 23.5 m → ~ USD 92 m projected by 2041)
	Marine spatial planning	Country-wide Marine Spatial Plan established
	Strengthened conservation governance	Independent Conservation Fund; transparent and participatory allocation system

Source The Nature Conservancy (2021)

2. **Biodiversity protection**: These schemes compensate land users for conserving or restoring habitats that support species diversity and ecological connectivity. *Example:* Conservation donors pay local communities to set aside or naturally restore land, creating biological corridors or protected areas.
3. **Watershed protection**: Payments target land-use practices that secure water quality and reduce hydrological risks. *Example:* downstream water users remunerate upstream farmers who adopt soil

conservation and reforestation practices that limit erosion, deforestation, and flooding.

4. **Landscape beauty and cultural services**: PES schemes can also address non-material ecosystem services, such as aesthetic and recreational values. *Example:* a tourism operator pays a local community not to hunt in a forest used for wildlife viewing, thus preserving the attractiveness and ecological integrity of the area.

These examples illustrate how PES schemes translate ecological functions into economic incentives, creating a market-like mechanism that rewards the provision and maintenance of nature's benefits.

In their traditional form, PES do not purely qualify as blended finance. Most existing PES programmes are publicly financed or supported by development institutions. Traditional public schemes, such as Costa Rica's *Pago por Servicios Ambientales* or China's *Sloping Land Conversion Program,* pioneered the model, compensating landholders for reforestation and watershed protection (Bennett, 2008; Pagiola, 2008). These early experiences demonstrated that targeted financial incentives can shift land-use decisions and generate measurable ecological benefits, while also reducing rural poverty.

However, the growing interest in market-based conservation has opened opportunities for private financial actors to play a complementary role. Banks and investors are not usually the direct funders of PES, but they can contribute through three functional channels: intermediation, co-financing, and innovation in financial products. In particular, banks can act as financial intermediaries within PES programmes, managing payments, ensuring transparency, and providing access to formal banking services for local beneficiaries. A notable example is *BanCO2* in Colombia, where Bancolombia manages the transfer of payments from corporate contributors to rural households that protect forests and water resources.[30] In this case, the bank's involvement has improved traceability, reduced transaction costs, and enabled the inclusion of local communities in the formal financial system. Furthermore, banks can contribute indirectly by integrating PES logic into sustainability-linked financial products. For instance, the outcomes generated by PES (verified carbon sequestration or watershed restoration) can serve as measurable indicators in green loans, sustainability-linked loans, or biodiversity credit portfolios. Although this integration is still

[30] https://www.cooperacionsuiza.pe/march-2017-8/.

emerging, it offers a pathway for banks to internalise ecosystem services into credit assessment and product design, linking lending conditions to nature-related performance.

5.4 Challenges and New Paths

The transition from risk management to nature-positive finance positions banks at the heart of the global effort to halt biodiversity loss. As this chapter has shown, financial institutions can influence both sides of the equation: reducing harmful financial flows through exclusion and engagement strategies, and mobilising new capital towards restoration through innovative credit, capital-market, and blended finance instruments. Together, these actions translate biodiversity from an externality into a financially material variable and a new frontier of strategic banking innovation.

Closing the biodiversity finance gap will require a systemic approach: integrating nature-related risk assessment into prudential regulation, standardising biodiversity metrics across products, and fostering public–private partnerships that enhance project bankability. Banks' ability to act as intermediaries between global investors, development institutions, and local conservation actors will determine whether biodiversity finance remains a niche or becomes a mainstream component of sustainable finance.

5.4.1 Barriers in Mobilising Private Finance

Yet, despite the growing attention to nature-positive finance, the mobilisation of private capital for biodiversity conservation still faces multiple structural barriers. As highlighted by Cosma et al. (2023), financing conservation projects implies confronting high uncertainty, long time horizons, elevated transaction costs, and the absence of predictable cash flows. We are talking about conditions that undermine the conventional criteria of financial decision-making based on risk, return, and value. These characteristics make biodiversity investments difficult to price and assess within traditional banking and asset-management frameworks.

Similarly, Sheikh et al. (2025) describe most nature-related projects as *at the frontier of bankability*: unproven, small-scale, and lacking sufficient revenue generation to meet private financiers' risk-return thresholds. The result is a persistent misalignment between the public good nature of biodiversity and the private incentives guiding capital allocation. Beverdam et al. (2025) further note that, because biodiversity seldom yields direct financial returns, private investors tend to avoid such projects unless adequate mechanisms exist to ensure liquidity, diversification, and return comparability with other asset classes. In practice, high monitoring and verification costs, illiquid secondary markets, and uncertain property-rights regimes discourage large-scale private engagement.

Sheikh et al. (2025) add that these financial obstacles intertwine with institutional, informational, and technical constraints. Fragmented and unstable regulatory frameworks, weak governance of natural resources, and the absence of consistent de-risking incentives prevent biodiversity finance from achieving predictable long-term conditions. At the same time, the lack of standardised data and metrics complicates the valuation of biodiversity-related assets and their integration into risk models. Many financial institutions still lack the internal expertise to evaluate ecological risks and performance, amplifying informational asymmetries and perceived uncertainty.

5.4.2 Don't Leave Private Finance Alone

Taken together, these barriers explain why private finance remains cautious and why public and blended instruments are still essential for bridging the gap between conservation needs and market expectations. Overcoming them will require innovative mechanisms for risk-sharing and incentive alignment, including first-loss guarantees, outcome-based payments, and fiscal measures that improve project bankability. Advancing research on portfolio selection, valuation methodologies, and cross-sectoral risk sharing is equally crucial to redefine biodiversity as an investable asset class rather than a philanthropic endeavour (Cosma et al., 2023).

In this regard, Garvey et al. (2025) emphasise that scaling up private investment in ecological restoration will not be achieved merely by "creating new bonds" or repackaging existing green instruments. Rather, it requires a dedicated financial architecture capable of:

- aggregating highly heterogeneous projects under common investment vehicles;
- managing non-traditional risks through hybrid risk-transfer mechanisms;
- improving transparency via robust geospatial data and verifiable impact metrics; and
- connecting *payors* (those who benefit from ecosystem services) with on-the-ground conservation projects.

The authors argue that private finance alone cannot solve the problem: the public sector remains indispensable, both as guarantor and as systemic coordinator. The true innovation, therefore, lies not in issuing isolated "green" products but in restructuring the financial ecosystem itself so that it becomes genuinely compatible with ecological and territorial dynamics.

Ultimately, biodiversity finance is not only about preserving nature; it is about redefining the financial system's capacity to generate value within planetary boundaries. Banks that embrace this paradigm shift, from financing economic growth *despite* nature to financing growth *through* nature, will not only mitigate their own risks but also shape the architecture of a truly regenerative economy. Academic research is called to demonstrate that this shift generates more value.

References

Agliardi, E., & Agliardi, R. (2025). Biodiversity-related sustainability-linked bonds. In T. Walker, H. Naffa, R. K. Tharumar, & S. Donders (Eds.), *Biodiversity finance. Palgrave studies in emerging risk management and sustainable finance*. Palgrave Macmillan.

Arjaliès, D. L., Bernard, J., & Patel, D. (2025). The potential of biodiversity offsets and credits to finance nature-based solutions. In *Biodiversity finance: The economic, operational, and societal impacts of biodiversity loss* (pp. 233–266). Springer.

Azizi, L., Scope, C., Ladusch, A., & Sassen, R. (2025). Biodiversity disclosure in the European finance sector. *Ecological Economics, 228*, Article 108430.

Bennett, M. T. (2008). China's sloping land conversion program: Institutional innovation or business as usual? *Ecological Economics, 65*(4), 699–711.

Berrada, T., Engelhardt, L., Gibson, R., & Krueger, P. (2022). The economics of sustainability-linked bonds. *Swiss Finance Institute Research Paper* (22–26).

Beverdam, J., Hubacek, K., Scholtens, B., & Sijtsma, F. (2025). Improving biodiversity resilience requires both public and private finance: A life-cycle analysis of biodiversity finance. *Ecological Economics, 234*, Article 108607.

Climate Bonds Initiative. (2023). *Green bond pricing in the primary market: H2 2022*. London: Climate Bonds Initiative. Available at: https://www.climatebonds.net/files/documents/publications/Green-Bond-Pricing-in-the-Primary-Market-H2-2022.pdf

Cojoianu, T. F., Ascui, F., Clark, G. L., Hoepner, A. G., & Wójcik, D. (2021). Does the fossil fuel divestment movement impact new oil and gas fundraising? *Journal of Economic Geography, 21*(1), 141–164.

Cosma, S., Galletta, S., Mazzù, S., & Rimo, G. (2024). Banks' fossil fuel divestment and corporate governance: The role of board gender diversity. *Energy Economics, 139*, Article 107948.

Cosma, S., Rimo, G., & Cosma, S. (2023). Conservation finance: What are we not doing? A review and research agenda. *Journal of Environmental Management, 336*, Article 117649.

Croci, E., Lucchitta, B., & Cusa, M. (2025). Biodiversity credits schemes: A comparative analysis. *Journal of Cleaner Production, 523*, Article 146382.

Deutz, A., Heal, G. M., Niu, R., Swanson, E., Townshend, T., Zhu, L., Delmar, A., Meghji, A., Sethi, S. A., & Tobin-de la Puente, J. (2020). Financing nature: Closing the global biodiversity financing gap. *The Paulson Institute, The Nature Conservancy, and the Cornell Atkinson Center for Sustainability, 256*.

Di Tommaso, C., Pacelli, V., & Povia, M. M. (2025). Green loans and bank risk: Navigating the path to sustainable finance. *International Review of Economics and Finance, 2025*, 104138.

Du, K., Harford, J., & Shin, D. D. (2023). Who benefits from sustainability-linked loans? *European Corporate Governance Institute–Finance Working Paper* (917).

Ducros, A., & Steele, P. (2022). *Biocredits to finance nature and people: emerging lessons*. International Institute for Environment and Development.

Elliott, J., & Löfgren, Å. (2022). If money talks, what is the banking industry saying about climate change? *Climate Policy, 22*(6), 743–753.

European Banking Authority. (2023). *EBA report in response to the call for advice from the European Commission on green loans and mortgages* (EBA/REP/2023/38). European Banking Authority.

Fenichel, E. P., & Dean, M. F. (2024). Blended academic insights for biodiversity and conservation finance. *Ecological Economics, 223*, Article 108258.

Flammer, C. (2021). Corporate green bonds. *Journal of Financial Economics, 142*(2), 499–516.

Flammer, C., Giroux, T., & Heal, G. M. (2025). Biodiversity finance. *Journal of Financial Economics, 164*, Article 103987.

Fresnillo, I. (2023). Miracle or mirage: Are debt swaps really a silver bullet? *Brussels: European Network on Debt and Development. December*. https://www.eurodad.org/miracle_or_mirage

Fuller, K. (1989). ES&T views: Debt-for-nature swaps. *Environmental Science and Technology, 23*(12), 1450–1451.

Garvey, J., Grigoriadis, V., Flannery, D., Knapp, E., Gold, E., Hutchinson, G., & Byrne, K. A. (2025). Designing financial instruments for land-based ecological restoration: A review and future research Agenda. *Cleaner Production Letters, 8*, 100089.

Harrer, T., Silvola, H., & Lehner, O. M. (2023). Measuring biodiversity: Mission impossible? In *The Routledge handbook of green finance* (pp. 357–373). Routledge.

International Bank for Reconstruction and Development. (2022). *Wildlife conservation bonds: Prospectus supplement (March 28, 2022)*. World Bank.

International Capital Market Association. (2023). *Green bond principles: Voluntary process guidelines for issuing green bonds*. ICMA. Available at: https://www.icmagroup.org/sustainable-finance/the-principles-guidelines-and-handbooks/

ItaSIF. (2024). *Sustainable finance and biodiversity—A guide for financial operators*. Available at: https://finanzasostenibile.it/en/attivita/sustainable-finance-and-biodiversity-a-guide-for-financial-actors/

Kedward, K., Zu Ermgassen, S., Ryan-Collins, J., & Wunder, S. (2023). Heavy reliance on private finance alone will not deliver conservation goals. *Nature Ecology and Evolution, 7*(9), 1339–1342.

Krueger, P., Sautner, Z., & Starks, L. T. (2020). The importance of climate risks for institutional investors. *The Review of Financial Studies, 33*(3), 1067–1111.

Larcker, D. F., & Watts, E. M. (2020). Where's the greenium? *Journal of Accounting and Economics, 69*(2–3), Article 101312.

Matthews, A., & Karousakis, K. (2022). Identifying and assessing subsidies and other incentives harmful to biodiversity: A comparative review of existing national-level assessments and insights for good practice. *OECD Environment Working Papers*, No. 206, OECD Publishing, Paris.

Mundaca, L., & Heintze, J. N. (2024). Banking on ecosystem services. *Ecological Economics, 224*, Article 108284.

Occhiolini, M. (1990). *Debt-for-nature swaps* (Vol. 393). World Bank Publications.

Pagiola, S. (2008). Payments for environmental services in Costa Rica. *Ecological Economics, 65*(4), 712–724.

Reghezza, A., Altunbas, Y., Marques-Ibanez, D., d'Acri, C. R., & Spaggiari, M. (2022). Do banks fuel climate change? *Journal of Financial Stability, 62*, Article 101049.

Sheikh, H. A., Narain, D., Bartlett, C., & Christiaen, C. (2025). Unlocking private finance for nature: Addressing barriers and reframing risk-return dynamics. *iScience, 28*(11), 1.

Simeth, N. (2025). Debt-for-nature swaps: A case study of Gabon. *Emerging Markets Review, 65*, Article 101244.

The Nature Conservancy. (2021). *Belize debt conversion for marine conservation: Case study*. The Nature Conservancy.

TNFD. (2023). Recommendations of the taskforce on nature-related financial disclosures. *Taskforce on Nature-related Financial Disclosures.*
Willis, J. (2026). Biodiversity finance: What has been achieved and where is it heading. *Ecological Economics, 240*, Article 108713.
World Bank. (2022). *South Africa–South African National Parks and the Eastern Cape Parks and Tourism Agency: Wildlife Conservation Bond Operation Project (P174097)* [Project Appraisal Document No. PIDC31569]. Washington, DC: World Bank Group. Available at https://documents1.worldbank.org/curated/en/295061649080092070/pdf/South-Africa-South-African-National-Parks-and-the-Eastern-Cape-Parks-and-Tourism-Agency-Wildlife-Conservation-Bond-Operation-Project.pdf
Wunder, S. (2005). *Payments for environmental services: Some nuts and bolts* (Vol. 42, pp. 1–32). Bogor.
Wunder, S., Fraccaroli, C., Bull, J. W., Dutta, T., Eyres, A., Evans, M. C., & zu Ermgassen, S. O. (2025). Biodiversity credits: An overview of the current state, future opportunities, and potential pitfalls. *Business Strategy and the Environment, 2025*, 1.

CHAPTER 6

The Effect of Biodiversity and Environmental Risks on Banks' Performance

Abstract This chapter presents an empirical analysis of how integrating Biodiversity and Environmental Risks (BER) into credit screening affects bank performance. Building on the theoretical foundations developed in earlier chapters and examining a global panel of 726 banks across 56 countries from 2015 to 2024 through fixed-effects panel regressions, we document three main findings. First, BER integration is associated with higher asset quality, reflected in significantly lower non-performing loan ratios. Second, our results show that banks that take biodiversity criteria into account have greater profitability in terms of ROA and ROE, which implies a better intermediation process and risk control. Third, BER integration seems to be valued by markets, as suggested by its positive and significant association with Tobin's Q. These results are not homogeneous: performance benefits emerge in non-megadiverse countries, where ecological risks seem to be less systemic and BER practices remain less homogenised. Finally, the Kunming Declaration, the Glasgow Climate Pact, and the Basel Committee's focus on climate risk seem to mark a turning point: the improvements resulting from BER integration in credit screening are reflected in accounting indicators only after 2021. Overall, the evidence shows that integrating biodiversity risks is not merely a compliance requirement, but a strategic and prudential tool that enhances banks' resilience in a context of accelerating environmental degradation.

G. Rimo and S. Cosma, *Biodiversity and Banks*, Palgrave Studies in Impact Finance, https://doi.org/10.1007/978-3-032-17631-8_6

Keywords Biodiversity finance · Biodiversity risk · Bank performance · Credit risk · Sustainable banking

6.1 Introduction

Risks arising from biodiversity loss and, more generally, environmental degradation are multifaceted, as analysed in depth in the previous chapters. These facets represent a real call to action for the banking sector, which must manage these risks that could amplify traditional prudential risks. Indeed, banks operate at the core of financial intermediation and are exposed to nature-related risks, mainly through their lending and investment activities. Banking cannot, of course, be considered a directly polluting activity or one that has an impact on soil, air, or water quality. However, as Thompson and Cowton (2004) point out, lending ties banks inextricably to the pollution generated by the commercial activities they finance.

Biodiversity-related risks can show up on a bank's balance sheet in several ways. Borrowers may be directly exposed to biodiversity loss, collateral can lose value when ecosystems deteriorate, new rules can tighten lending conditions, and environmental shocks can even ripple through the system more broadly. From a credit-risk angle, banks can end up taking losses when the firms they lend to are hit by biodiversity degradation—maybe because their operations are disrupted, natural inputs become less productive, remediation gets more expensive, or they are pushed towards cleaner (and sometimes costlier) production models (Hummel & Jobst, 2024; Mulder & Koellner, 2011).

There is also the less "visible" side of the problem. Banks can run into legal or reputational trouble if they are linked to nature-related controversies—for instance, when they finance companies involved in deforestation, ecosystem destruction, land-use conversion, or similar harmful practices (Adeabah et al., 2023; Boyer & Laffont, 1997). Even if the financial exposure isn't immediate, being associated with damaging activities can quickly become a liability. These risks, if not adequately monitored and priced, can impair loan quality, increase non-performing exposures, and undermine profitability.

Against this backdrop, supervisors have started to acknowledge that weaving Biodiversity and Environmental Risks (BER) into the credit-screening process isn't just a nice extra step, but it may actually become an important way for banks to reduce their exposure to nature-related vulnerabilities (Basel Committee, 2023; European Banking Authority, 2024). For example, the latest EBA guidelines on ESG risk management emphasise the need for institutions to integrate ESG risks into all traditional categories of financial risk. ESG factors (including those linked to ecosystem damage and biodiversity loss) aren't meant to sit in a separate box or be treated as an add-on. They need to be woven directly into a bank's risk management framework: its strategies, policies, and risk limits. Within the set of tools highlighted by the EBA, one element stands out: active engagement with counterparties. Banks are expected to deepen their understanding of a counterparty's risk profile, evaluate how resilient it is to ESG-related pressures, and, when necessary, agree on concrete actions tailored to that counterparty. These steps can include adjusting the terms of a product, agreeing on transition or remediation plans, or putting in place measures that make the counterparty more capable of handling ESG-related risks. It's only when these attempts don't work, and when keeping the relationship would clash with the bank's strategic priorities or risk appetite, that the EBA views ending the relationship as an acceptable course of action. In other words, divestment from counterparties that are heavily exposed to ESG risks should be viewed strictly as a last resort (European Banking Authority, 2025).

The academic literature offers only partial guidance on this field. Over the years, numerous scholars have investigated the relationship between more general ESG or Corporate Social Responsibility (CSR) practices and the financial performance of banks, yielding mixed results that have fuelled two distinct views within the literature.

The first perspective, which is based on the *resource-based view* and *stakeholder theory*, considers ESG and CSR initiatives as strategic activities (Donaldson & Preston, 1995; Freeman, 1984; Russo & Fouts, 1997). By strengthening their competitive advantage and eventually improving their long-term financial performance, these practices help businesses build distinctive resources and capabilities. At the same time, they allow firms to fulfil their moral duty towards the broader network of stakeholders in which they operate by addressing their needs and interests, thereby reinforcing the firm's legitimacy within its institutional environment.

The second view, based on *shareholder theory*, regards the company as an entity whose sole moral obligation is to maximise shareholder profits (Friedman, 2007; Smith & Rönnegard, 2016). This view, therefore, considers ESG or CSR activities to be an inefficient allocation of capital.

Empirical evidence has not provided clear-cut results that would allow us to consider only one of these theories to be correct. Wu and Shen (2013), for example, found evidence that banks' investments in CSR activities can lead to improved financial performance, as measured by return on assets, return on equity, and net interest income, as well as a reduction in non-performing loans. Cantero-Saiz et al. (2024) and Liu et al. (2023) support this view, highlighting a positive relationship between ESG and asset quality. Conversely, Scholtens and Dam (2007) point out that banks adhering to the Equator Principles have higher CSR policies but, at the same time, lower returns on assets. Similarly, Agnese et al. (2024) note that banks with a less focused approach to ESG are also those with better financial performance.

Azmi et al. (2021) highlight a non-linear relationship between ESG activity and bank value. In particular, the authors find that low levels of ESG activity lead to a higher Tobin's Q. A similar view is supported by the study by El Khoury et al. (2023), according to which incremental ESG investments bring benefits to the profitability and value of banks up to a certain inflexion point, beyond which this relationship is reversed.

Citterio and King (2023) analyse the predictive power of ESG indicators to forecast bank financial distress, highlighting how these factors can significantly reduce the likelihood of misclassifying distressed/defaulted banks as healthy. Cao et al. (2024) also demonstrate that greater ESG investments can increase banking efficiency. Similarly, Chiaramonte et al. (2022) find evidence of a stabilising effect of banking ESG strategies during periods of financial turmoil, while Galletta et al. (2023) suggest that such strategies can reduce reputational risk. On the other hand, Di Tommaso and Thornton (2020) suggest that, although the implementation of ESG activities can slightly reduce bank risk-taking, it also leads to a reduction in bank value.

Finally, within the extensive ESG literature, there is a strand of research examining the impact of climate change exposure on banks' financial stability. For example, Li and Pan (2022) found evidence of a detrimental effect of climate transition risk on banks' financial performance, partly by reducing the scale of bank loans. Chalabi-Jabado and Ziane (2024) show that climate transition risk and physical climate risk don't exert the same effect on banks' performance. Transition risk actually seems to support performance and loan growth, suggesting that climate policies can work in banks' favour. Physical risk, instead, tends to move in the opposite direction, hurting both profitability and lending. Birindelli et al. (2022) report a negative link between banks' commitment to climate issues and the overall riskiness of their loan portfolios. Alessi et al. (2024) add another layer to the picture: they show that transition risk can generate sizable losses for the banking sector and, in the most severe scenarios, could even push some institutions towards default.

Therefore, studies on ESG and banking performance suggest that environmental practices can strengthen asset quality and reduce credit risk, although sometimes with negative impacts on profitability. However, research explicitly linking biodiversity, as opposed to broader environmental indicators, to banking outcomes remains extremely limited.

While previous chapters discussed the theoretical channels through which biodiversity loss affects financial stability, a key empirical question remains open: *Do banks that explicitly integrate biodiversity and environment considerations into their lending criteria perform better than those that do not?* Moreover, if so, through which dimensions of performance (credit quality, profitability, or market valuation) do these benefits materialise?

By basing the theoretical arguments developed throughout the previous sections on systematic empirical evidence, this chapter demonstrates how integrating biodiversity risk can serve as a strategic and prudential tool for improving resilience in an era of accelerating environmental degradation.

6.2 Data and Methodology

The empirical analysis relies on a panel dataset covering 726 global banks from 2015 to 2024 (hence for 10 years) and includes financial, structural, and governance-related variables. The sample was first identified by selecting all financial companies belonging to the banking sector on LSEG Data & Analytics (formerly Refinitiv Eikon) using the Global Industry Classification Standard (GICS). We then employed a sample cleaning procedure that resulted in the exclusion of all banks that lacked data related to our main variables of interest for at least 1 year.

To test our research idea, and thus to study the effect of BER integration on banking performance, we estimate the following fixed-effects panel regression model:

$$BankPerformance_{it} = \alpha_i + \beta BER_{it} + \theta' Financial_{it} + \delta' Board_{it} + \varepsilon_{it}; \quad i = 1, 2, \ldots, N; \quad t = 1, 2, \ldots, T$$

where i indexes banks, t indexes years and $\varepsilon_{i,t}$ is the error term. Furthermore, we control for time-varying, country-specific shocks by including country-year fixed effects (Berger et al., 2024; Tan et al., 2021).

To measure banking performance (*BankPerformance*) from various perspectives, we use a series of dependent variables commonly employed in the financial literature. Specifically, the *NPL ratio* (non-performing loans to total loans) represents a key credit quality measure with respect to banks' lending practices, serving as a proxy for credit risk and asset quality (Liang et al., 2013; Liu et al., 2023). *ROA* (pretax income to total assets) is a measure of bank profitability widely used in the literature and allows us to measure how efficiently the bank uses its assets to generate revenues (García-Meca et al., 2015; Liang et al., 2013), while *ROE* (pretax income to total equity) reflects the financial return generated on shareholders' invested capital (Liang et al., 2013). As a market-based measure, we employ Tobin's Q (Ln(Q)), computed as the ratio of market capitalisation to the replacement cost of total assets in its logarithmic form (Aouadi & Marsat, 2018). This measure signals future profitability and reflects unstructured and volatile bank performance (Azmi et al., 2021; García-Meca et al., 2015).

The key independent variable is *BER*, a dummy equal to one if the bank explicitly states that it considers biodiversity- and environment-related risks when evaluating projects to be financed, and zero otherwise. This indicator reflects the integration of environmental risk assessment into credit decision-making processes and serves as a proxy for the bank's awareness of biodiversity and sustainability orientation.

To control for bank heterogeneity, several financial and governance control variables are included, generally used in the most authoritative banking literature (Azmi et al., 2021; García-Meca et al., 2015; Liang et al., 2013). *LnTA* (the natural logarithm of total assets) captures size effects; *ETA* (equity to total assets) represents capitalisation; *Liquidity* (deposits to total assets) measures the funding structure; and *LoanRatio* (loans to total assets) indicates the degree of credit intermediation.

Tier 1 Capital accounts for the regulatory capital position. The *BusinessMod* variable, defined as non-interest income over total revenue, captures the diversification of income sources, while *EfficiencyRatio* (cost-to-income ratio) reflects managerial efficiency. Board characteristics are also considered (Liang et al., 2013): *BoardSize* (number of directors), *BoardGender* (female-to-total board ratio), and *IndependentBoard* (independent-to-total directors ratio) control for governance quality and potential heterogeneity in board composition.

All continuous variables are winsorized to reduce the impact of extreme observations and ensure that outliers do not significantly influence the empirical results. In particular, in accordance with standard practice in empirical banking and corporate finance research, we apply a symmetric winsorization at the 1st and 99th percentiles (Bhagavatula et al., 2023; Cumming et al., 2025). This process reduces the leverage of potentially anomalous values, such as abnormally high profitability ratios or capitalisation levels, while maintaining the distributional properties of the data.

The list of variables used, along with their definitions, is provided in Table 6.1.

6.3 Descriptive Statistics

Table 6.2 reports our descriptive statistics. Our bank sample displays substantial heterogeneity in performance, balance sheet structure, and governance characteristics. In particular, credit-risk levels are moderate: the average NPL ratio is 2.9%, with relatively low dispersion, although a few institutions exhibit higher levels of impaired loans (up to 35.6%). The LoanRatio, at 63% on average, highlights a business model strongly oriented towards traditional intermediation, while the EfficiencyRatio highlights notable variation in managerial efficiency. Governance characteristics also vary considerably across institutions. Boards have an average size of 11 members, with female representation around 19% and independent directors accounting for 61% of the board.

According to our summary statistics, the primary explanatory variable, BER, equals one for approximately 48% of the sample, suggesting that nearly half of the banks explicitly integrate biodiversity and environmental-related risks into their credit evaluation processes. Figure 6.1 plots the annual percentage of banks that explicitly integrate biodiversity and environmental-related risks into their credit-risk

Table 6.1 Variables employed in the analysis and definitions

Variable	*Definition*
Dependent variables	
NPL ratio	Non-performing loans on total loans
ROA	Pretax income on total assets
ROE	Pretax income on total equity
Ln(*Q*)	Ratio of the market capitalisation to the replacement cost of its total assets. Implemented in its natural logarithm form
Independent variable	
BER	Dummy variable that takes the value of one if the bank claims to consider biodiversity and environmental risks when evaluating projects to be financed, and 0 otherwise
Control variables	
LnTA	Logarithm of total assets
ETA	Total equity on total assets
Liquidity	Total deposits on total assets
LoanRatio	Total loans on total assets
Tier1Capital	Tier 1 Capital on total risk-weighted assets
BusinessMod	Non-interest income on total revenue
EfficiencyRatio	Cost-income ratio
BoardSize	Number of board members
BoardGender	Ratio of female directors to the total number of board members
IndependentBoard	Ratio of independent directors to the total number of board members

Note The table provides definitions for the dependent variables, the main independent variable, and the control variables used in the analysis. Each variable is sourced from LSEG Data and Analytics (formerly Refinitiv Eikon). *Source* Authors' elaboration

assessment processes. This graph highlights a structural acceleration in banks' biodiversity awareness after 2019. This sustained rise is consistent with the broader post-2020 policy momentum around nature-related financial risks, including the release of international frameworks such as the Kunming Declaration, the growing prominence of TNFD, and the integration of biodiversity criteria into European sustainable finance regulations.

Table 6.2 Descriptive statistics

Variable	*Mean*	*Std. Dev*	*Min*	*Max*
NPLRatio	0.029	0.047	0	0.356
ROA	0.014	0.012	−0.021	0.066
ROE	0.135	0.094	−0.211	0.49
Ln(Q)	−2.346	0.767	−4.445	−0.268
BER	0.479	0.5	0	1
LnTA	23.803	1.918	20.25	28.592
ETA	0.104	0.04	0.04	0.304
Liquidity	0.747	0.117	0.31	0.908
LoanRatio	0.633	0.144	0.161	0.89
Tier1Capital	0.145	0.042	0.086	0.347
BusinessMod	0.313	0.161	0.023	0.804
EfficiencyRatio	0.587	0.154	0.26	1.157
BoardSize	11.422	3.229	5	21
BoardGender	19.411	13.691	0	55.556
IndependentBoard	61.435	25.976	0	100

Note The table provides descriptive statistics (mean, standard deviation, minimum, and maximum) for the variables used in the empirical analyses. *Source* Authors' elaboration

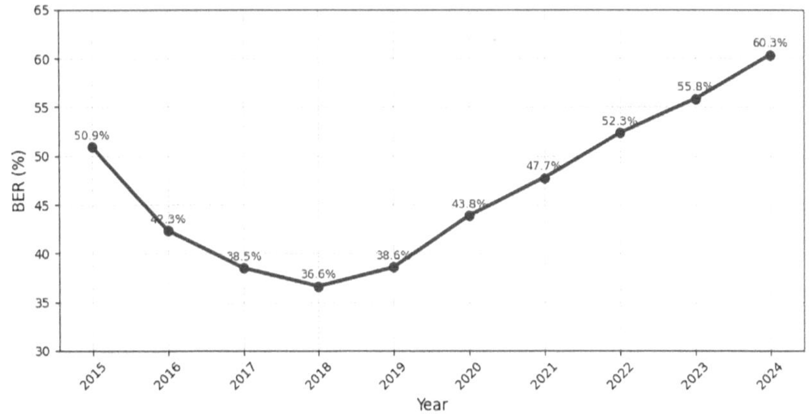

Fig. 6.1 Banks evaluating biodiversity and environmental risks (BER), 2015–2024. *Source* Authors' elaboration

6.4 Empirical Results

Table 6.3 reports the baseline fixed-effects estimations. Across all specifications, the coefficient on BER is statistically significant and displays a pattern consistent with enhanced asset quality, improved profitability, and higher market valuation.

In column 1, our results indicate that banks that explicitly integrate biodiversity and environmental considerations into their credit screening

Table 6.3 Baseline results

	(1) *NPLRatio*	*(2)* *ROA*	*(3)* *ROE*	*(4)* *Ln(Q)*
BER	−0.0039***	0.0005*	0.0074***	0.0657***
	(0.0015)	(0.0003)	(0.0022)	(0.0219)
lnTA	−0.0011**	−0.0006***	−0.0068***	−0.0377***
	(0.0005)	(0.0001)	(0.0007)	(0.0084)
ETA	0.1125***	0.0633***	−0.5692***	6.6461***
	(0.0281)	(0.0061)	(0.0421)	(0.4515)
Liquidity	−0.0122	0.0062***	0.0435***	0.6949***
	(0.0091)	(0.0016)	(0.0125)	(0.1174)
LoanRatio	−0.0421***	0.0023**	0.0026	−0.0123
	(0.0126)	(0.0010)	(0.0093)	(0.0973)
Tier1Capital	−0.0819**	0.0095*	0.0973**	2.4532***
	(0.0415)	(0.0054)	(0.0396)	(0.4536)
BusinessMod	−0.0212***	0.0175***	0.1594***	0.3164***
	(0.0064)	(0.0012)	(0.0094)	(0.0758)
EfficiencyRatio	0.0175***	−0.0406***	−0.3924***	−1.3053***
	(0.0066)	(0.0011)	(0.0098)	(0.0741)
BoardSize	−0.0004*	−0.0000	−0.0005**	0.0071***
	(0.0002)	(0.0000)	(0.0003)	(0.0027)
BoardGender	−0.0000	0.0000***	0.0002***	0.0028***
	(0.0001)	(0.0000)	(0.0001)	(0.0007)
IndependentBoard	−0.0001**	0.0000**	0.0002**	0.0015**
	(0.0000)	(0.0000)	(0.0001)	(0.0006)
Constant	0.0977***	0.0312***	0.4766***	−2.5740***
	(0.0266)	(0.0032)	(0.0259)	(0.2769)
Observations	4166	4236	4236	4229
Country × year FE	Yes	Yes	Yes	Yes
R^2	0.5880	0.7678	0.7770	0.7446

Note The table shows the fixed-effects regression results on the effect of BER on NPL Ratio (Column 1), ROA (Column 2), ROE (Column 3), and Tobin's Q (Column 4). All variables are defined in Table 6.1. Standard errors in parentheses

*p < 0.10, **p < 0.05, ***p < 0.01

exhibit a lower NPL ratio. The negative and highly significant coefficient suggests that incorporating nature-related risks helps banks avoid borrowers with hidden exposure to environmental liabilities, ecosystem degradation, or regulatory pressure. This result is consistent with BER acting as an additional layer of due diligence that improves loan selection and reduces the likelihood of portfolio deterioration. This result aligns with the growing empirical evidence showing that stronger ESG practices in the banking sector are associated with lower credit risk and improved asset quality, typically reflected in reduced non-performing loans (Baek & Kang, 2025; Cantero-Saiz et al., 2024; Liu et al., 2023).

Turning to profitability, BER is positively associated with both ROA and ROE (columns 2 and 3), indicating that banks adopting biodiversity-risk integration benefit from more efficient credit allocation and reduced loss provisioning needs. The improvement in accounting profitability aligns with the idea that better anticipation of environmental risks stabilises earnings and enhances risk-adjusted returns. Moreover, column 4 shows a strong and positive relationship between BER and Tobin's Q, indicating that equity markets systematically reward banks perceived as more forward-looking in managing emerging BER. This result suggests that investors interpret BER as a credible signal of superior governance quality, lower exposure to future environmental shocks, and better alignment with evolving regulatory expectations. These relationships are consistent with the broader literature, which suggests that enhanced environmental practices can strengthen bank performance by mitigating risk and improving operational resilience (Azmi et al., 2021).

Taken together, these baseline results provide consistent evidence that integrating biodiversity and environmental risks into credit decision-making is associated with improved performance across asset quality, profitability, and market valuation. This pattern reinforces the idea that BER integration is not merely a compliance-driven exercise, but a strategically valuable practice that enhances banks' resilience in the face of rising nature-related risks.

6.4.1 Robustness Check

To address some endogeneity or simultaneity issues, we re-estimated the baseline specifications from Table 6.3 by applying a 1-year time lag to all independent variables (Bigelli et al., 2023). The results in Table 6.4 show that the coefficient relating to lagged BER remains statistically significant in all the specifications adopted and thus confirm our basic findings. The stability of the estimates reinforces the robustness of our findings and reduces concerns that reverse causality or simultaneous determination drive the associations observed in the baseline models.

Table 6.4 Models with lagged independent variables

	(1) *NPLRatio*	(2) *ROA*	(3) *ROE*	(4) *Ln*(Q)
L.BER	−0.0033**	0.0006*	0.0072**	0.0720***
	(0.0016)	(0.0004)	(0.0032)	(0.0245)
L.lnTA	−0.0015***	−0.0006***	−0.0047***	−0.0442***
	(0.0005)	(0.0001)	(0.0010)	(0.0091)
L.ETA	0.1207***	0.0702***	−0.4709***	6.3412***
	(0.0365)	(0.0079)	(0.0581)	(0.4783)
L.Liquidity	−0.0134	0.0083***	0.0582***	0.6353***
	(0.0102)	(0.0019)	(0.0154)	(0.1183)
L.LoanRatio	−0.0462***	−0.0010	−0.0094	−0.1471
	(0.0137)	(0.0015)	(0.0118)	(0.1055)
L.Tier1Capital	−0.0783*	0.0008	0.0001	2.0510***
	(0.0454)	(0.0068)	(0.0603)	(0.5082)
L.BusinessMod	−0.0222***	0.0138***	0.1142***	0.3963***
	(0.0072)	(0.0014)	(0.0119)	(0.0809)
L.EfficiencyRatio	0.0107*	−0.0295***	−0.2738***	−1.3682***
	(0.0057)	(0.0015)	(0.0130)	(0.0841)
L.BoardSize	−0.0003	−0.0000	−0.0009**	0.0064**
	(0.0002)	(0.0000)	(0.0004)	(0.0029)
L.BoardGender	−0.0000	0.0000*	0.0002*	0.0029***
	(0.0001)	(0.0000)	(0.0001)	(0.0008)
L.IndependentBoard	−0.0001*	0.0000***	0.0001	0.0012*
	(0.0001)	(0.0000)	(0.0001)	(0.0006)
Constant	0.1081***	0.0266***	0.3796***	−2.1803***
	(0.0284)	(0.0041)	(0.0337)	(0.2830)
Observations	3444	3771	3663	3664
Country × year FE	Yes	Yes	Yes	Yes
R^2	0.5823	0.5588	0.6340	0.7348

Note The table shows the fixed-effects regression results on the effect of BER on NPL Ratio (Column 1), ROA (Column 2), ROE (Column 3), and Tobin's Q (Column 4). A 1-year time lag is applied to the independent variables. All variables are defined in Table 6.1. Standard errors in parentheses *$p < 0.10$, **$p < 0.05$, ***$p < 0.01$

6.5 Additional Analysis

6.5.1 Countries' Biodiversity Richness

To examine whether the relationship between biodiversity-risk integration and bank performance varies across different ecological contexts, we perform an additional sample split based on countries' biodiversity richness. In particular, we distinguish between Megadiverse Countries, i.e., nations hosting more than 70% of the world's species, and non-megadiverse countries, which exhibit lower levels of biodiversity (Skoulodis et al., 2019; Cosma et al., 2025). This distinction is analytically relevant because banks operating in megadiverse jurisdictions are structurally more exposed to nature-related physical risks (e.g., deforestation, land-use change, ecosystem degradation), and often subject to more stringent or targeted environmental regulations. Conversely, in non-megadiverse economies, biodiversity risks tend to be more localised, idiosyncratic, and less embedded into national supervisory frameworks.

By splitting the sample along this dimension, we can assess whether differences in the materiality, regulatory salience, and systemic nature of biodiversity loss across countries drive the performance effects of integrating BER. In particular, this approach allows us to test whether integrating biodiversity considerations produces comparable benefits in contexts where ecological risks are widespread and largely non-diversifiable, compared to contexts where such risks are less pronounced and banks can enjoy a greater competitive advantage by adopting innovative and advanced environmental due diligence practices. Table 6.5 presents the results of this sample split.

Our results reveal a striking cross-country asymmetry: the integration of BER into credit screening is positively associated with bank performance in *non-megadiverse* countries (Panel B), whereas the effect becomes statistically insignificant in *megadiverse* countries (Panel A). This heterogeneity is theoretically consistent with differences in regulatory environments, market structure, and the underlying distribution of nature-related risks.

First, biodiversity risks are relatively less widespread and often underregulated in economies that are not megadiverse. As a result, only a small percentage of banks consider biodiversity, creating significant cross-sectional variation. By avoiding counterparties exposed to environmental liabilities, controversies, and ecosystem degradation, banks that implement rigorous environmental due diligence frameworks typically achieve

Table 6.5 Sample split—the role of countries' biodiversity richness

	Panel A: Megadiverse countries			
	(1) *NPLRatio*	*(2)* *ROA*	*(3)* *ROE*	*(4)* *Ln(Q)*
BER	−0.0003	−0.0001	0.0042	0.0212
	(0.0014)	(0.0003)	(0.0026)	(0.0305)
Observations	2589	2610	2610	2606
Controls	Yes	Yes	Yes	Yes
Country × year FE	Yes	Yes	Yes	Yes
R^2	0.3570	0.7206	0.7207	0.6847
	Panel B: Non-Megadiverse countries			
	(1) *NPLRatio*	*(2)* *ROA*	*(3)* *ROE*	*(4)* *Tobin's Q*
BER	−0.0084***	0.0010**	0.0085**	0.1118***
	(0.0027)	(0.0004)	(0.0037)	(0.0334)
Observations	1577	1626	1626	1623
Controls	Yes	Yes	Yes	Yes
Country × year FE	Yes	Yes	Yes	Yes
R^2	0.6076	0.8143	0.8231	0.7706

Note The table shows the fixed-effects regression results on the effect of BER on NPL Ratio (Column 1), ROA (Column 2), ROE (Column 3), and Tobin's Q (Column 4) for subsample of banks identified according to their belonging to Megadiverse countries. In Panel A, banks belong to Megadiverse countries. In Panel B, banks belong to non-Megadiverse countries. All variables are defined in Table 6.1. Standard errors in parentheses
*p < 0.10, **p < 0.05, ***p < 0.01

benefits in terms of credit allocation. This competitive edge reduces capital volatility and credit losses, ultimately enhancing market value and profitability. Put differently, in markets where peers have not yet internalised these risks, integrating biodiversity risks serves as a strategic differentiator.

On the other hand, the integration of environmental and biodiversity risks is likely more common in megadiverse economies, which are characterised by high levels of ecological richness, concentrated ecological pressures, and a significant reliance on natural assets. As a result, the institutional variance in risk-integration practices may be minimal, reducing the explanatory power of biodiversity-risk integration for performance outcomes. Statistically, when a practice becomes nearly universal, it cannot meaningfully predict performance differentials.

Moreover, in megadiverse countries, the physical dimension of biodiversity loss is inherently systemic, not idiosyncratic. A substantial share of economic activity is concentrated in sectors that are structurally vulnerable to ecosystem degradation, including agriculture, extractives, forestry, fisheries, and infrastructure. Thus, even when banks incorporate biodiversity-related criteria into credit processes, the baseline exposure of their loan portfolios remains high and largely non-diversifiable. Systemic nature-related risks cannot be mitigated through marginal adjustments in screening practices alone (Coqueret et al., 2025; Garel et al., 2024). As a consequence, the performance-enhancing benefits of biodiversity-risk integration become muted: integration is necessary, but not performance-differentiating.

Finally, in non-megadiverse countries, the incorporation of BER also acts as a credible governance signal, improving perceptions of risk management quality among investors and supervisors. In megadiverse countries, however, where environmental constraints are already priced into the macro-institutional framework, such signalling effects tend to be weaker.

Together, these mechanisms explain why the integration of biodiversity and environmental risks enhances performance only in non-megadiverse contexts: the practice is rare, differentiating, and effective in reducing idiosyncratic credit risk. In megadiverse countries, by contrast, it is common, institutionalised, and insufficient to overcome systemic ecological exposure, resulting in an overall neutral effect on performance.

6.5.2 *Event-Study*

To determine whether the integration of biodiversity and environmental risk influences bank performance differently before and after the intensification of global policy attention to nature-related risks, we employ a symmetric event-study approach centred on 2021. This was a particularly significant year for integrating BER into financial frameworks. First and foremost, it was the year of the Kunming Declaration (Garel et al., 2024). Signed in October 2021 during the first part of the UN Biodiversity Conference (COP15), the Declaration represents a significant international policy signal, calling for the alignment of financial flows with biodiversity objectives and urging financial institutions to strengthen their management of nature-related risks. 2021 also marked important steps forward for the climate transition. The COP26 in Glasgow marked a historic turning point in global climate governance by introducing, for the first time in a UN climate agreement, an explicit reference to coal. The Glasgow Climate Pact calls for the "phase-down of unabated coal power", signalling a collective commitment to reduce reliance on coal not equipped with emission-abatement technologies in a progressive manner. Major international banks pledged to effectively halt all international public financing for new unabated coal power projects by the end of 2021.[1] Moreover, in 2021, the Basel Committee on Banking Supervision published the "*Climate-related risk drivers and their transmission channels*", shedding light on the transmission channels through which climate-related risk drivers, including physical risks and transition risks, can affect the banking system.

We construct two symmetric sub-periods: a first window covering the years 2017–2020 and a second window covering the years 2021–2024, ensuring an equal number of years on each side of the event. This split enables us to examine whether increased policy attention to BER has acted as a catalyst for changes in banks' risk-governance practices. Table 6.6 presents the results of this symmetric event-study specification.

The estimates reveal an apparent asymmetry between accounting-based and market-based performance measures. For NPL ratio, ROA, and ROE, the coefficients associated with the post-2021 period are consistently significant and display the expected signs (Panel B). At the same time, those for the pre-2021 years are statistically insignificant (Panel A). This pattern suggests that improvements in loan portfolio quality and profitability emerge only after 2021, when global policy signals (including the Kunming Declaration, the Glasgow Climate Pact, and the BCBS report on climate-related risk drivers) appear to have accelerated the integration of BER into credit screening and internal risk management frameworks. Before 2021, BER was far less systematically factored into lending decisions, which explains the absence of performance effects in the pre-event window.

By contrast, Tobin's Q exhibits a markedly different dynamic. The valuation effect associated with biodiversity-risk integration is significant

[1] https://unfccc.int/news/end-of-coal-in-sight-at-cop26.

Table 6.6 The role of increased policy attention to BER in 2021

	Panel A: 2017–2020			
	(1) *NPLRatio*	*(2)* *ROA*	*(3)* *ROE*	*(4)* *Ln(Q)*
BER	−0.0035	0.0000	0.0032	0.1106***
	(0.0030)	(0.0004)	(0.0032)	(0.0388)
Observations	1612	1639	1639	1637
Controls	Yes	Yes	Yes	Yes
Country × year FE	Yes	Yes	Yes	Yes
R^2	0.6462	0.7774	0.7517	0.7807
	Panel B: 2021–2024			
	(1) *NPLRatio*	*(2)* *ROA*	*(3)* *ROE*	*(4)* *Ln(Q)*
BER	−0.0041**	0.0008*	0.0097***	0.0563**
	(0.0018)	(0.0004)	(0.0031)	(0.0285)
Observations	2108	2137	2137	2136
Controls	Yes	Yes	Yes	Yes
Country × year FE	Yes	Yes	Yes	Yes
R^2	0.4215	0.7619	0.7847	0.6924

Note The table shows the fixed-effects regression results on the effect of BER on NPL Ratio (Column 1), ROA (Column 2), ROE (Column 3), and Tobin's Q (Column 4) for two different sub-periods. Panel A covers the years 2017–2020, while Panel B covers the years 2021–2024. Standard errors in parentheses

$^{*}p < 0.10$, $^{**}p < 0.05$, $^{***}p < 0.01$

in both the pre- and post-2021 periods, with similar magnitudes across the two windows. This suggests that equity markets had already anticipated the increasing financial materiality of biodiversity loss and the likelihood of future regulatory tightening, even before the set of major policy signals was issued in 2021. Unlike accounting indicators, which capture realised outcomes, Tobin's Q reflects forward-looking expectations; investors rewarded banks perceived as more resilient to emerging BER long before these policy developments materialised.

Overall, the findings highlight a sharp divergence between ex ante market expectations and ex post realised performance. While market valuations responded smoothly across the entire event window, operational outcomes such as NPL, ROA, and ROE improved only after 2021, consistent with the view that the intensification of global policy attention in that year acted as a pivotal catalyst for accelerating biodiversity-risk integration practices within the banking sector.

6.6 Conclusions

This chapter provides the first empirical evidence on the relationship between biodiversity and environmental risk integration (BER) and bank performance, using a global panel of 726 banks from 2015 to 2024. The results consistently show that adding BER to credit screening is linked to better performance outcomes.

Firstly, BER integration appears to result in lower credit risk, which is reflected in a reduction in the NPL ratio. This result supports the idea that integrating nature-related criteria improves borrower selection and mitigates exposure to environmental liabilities. This conclusion is consistent with the growing body of research on ESG in banking, which shows how more stringent environmental and social policies enhance asset quality and lower non-performing loan rates (Baek & Kang, 2025; Cantero-Saiz et al., 2024; Liu et al., 2023). Second, there is a positive correlation between BER and ROA and ROE, suggesting that incorporating biodiversity risk encourages more effective and resilient intermediation. Third, BER is linked to a much higher Tobin's Q, indicating that equity markets reward banks that are proactive in handling new risks related to nature.

Significant contextual heterogeneity is also revealed by the analyses. Most of BER's performance benefits are found in non-megadiverse countries, where biodiversity risks are less common and regulatory requirements are more inconsistent. Adopting BER in these situations is a differentiating feature that lowers idiosyncratic risk and improves screening quality. We find no significant correlation between BER and performance in megadiverse countries, where environmental exposure is systemic and BER practices are more uniformly institutionalised. This disparity highlights that the performance benefits of biodiversity-risk integration are primarily found in areas where it is not yet widespread and where the underlying ecological risks are more variable.

The event study surrounding the increased policy attention to BER recorded in 2021 also shows that market expectations and bank behaviour were influenced by the momentum of nature-related policies. Both before and after 2021, market valuations responded to BER; however, improvements in accounting performance did not materialise until 2021, when global policy attention increased and biodiversity and environmental risk integration became more operationally embedded in risk management frameworks.

These results align with the growing awareness among central banks and financial regulators of the risks associated with the importance of nature to the economy. Authorities around the world are increasingly concerned about the potential financial instability and systemic risk that could result from the loss of biodiversity and ecosystem degradation, both through environmental damage ("physical risk") and through regulatory or policy responses to that damage ("transition risk"). As supervisory bodies, such as the Basel Committee (2023) and the European Banking Authority (2024), begin to examine these areas, financial institutions are being urged to assess the BER associated with the activities they fund and incorporate these factors into their risk management plans. Our findings suggest that these regulatory expectations are not merely compliance burdens; they may actually enhance banks' resilience and performance, particularly in jurisdictions where BER practices are still in development.

Our findings also carry important implications for the broader academic debate. A growing body of firm-level evidence suggests that biodiversity loss is associated with weaker operational performance, increased operational disruptions, and ultimately lower profitability (Ahmad & Karpuz, 2024; Cosma et al., 2025). By documenting that BER integration is associated with lower NPL ratios and stronger bank performance, our study extends this literature upstream along the financial intermediation chain, showing that the consequences of biodiversity degradation are not confined to firms but propagate to lenders. Moreover, the significant and forward-looking market response we observe is consistent with emerging evidence that investors are beginning to price a "biodiversity premium" (Giglio et al., 2025; Garel et al, 2024), rewarding institutions perceived as more resilient to nature-related risks. In this sense, our results reinforce the view that biodiversity is becoming financially material for the banking sector and increasingly relevant for both supervisors and market participants.

The evidence collectively indicates that integrating BER into risk management can increase bank resilience by enhancing credit screening, stabilising profitability, and strengthening market perception. However, it does not work for all banks; it depends on the rules and the environment in which they work. These findings suggest that enhancing biodiversity governance, supervisory expectations, and risk assessment capabilities can foster both prudential soundness and market competitiveness, particularly in jurisdictions where such practices are still in their early stages of development. Simultaneously, in contexts characterised by systemic and pervasive biodiversity loss, BER integration must be augmented by comprehensive structural and policy interventions to mitigate banks' nature-related vulnerabilities significantly.

References

Adeabah, D., Andoh, C., Asongu, S., & Gemegah, A. (2023). Reputational risks in banks: A review of research themes, frameworks, methods, and future research directions. *Journal of Economic Surveys, 37*(2), 321–350.

Agnese, P., Cerciello, M., Oriani, R., & Taddeo, S. (2024). ESG controversies and profitability in the European banking sector. *Finance Research Letters, 61*, Article 105042.

Ahmad, M. F., & Karpuz, A. (2024). Beyond climate change risk: Biodiversity and corporate cash holdings. *Economics Letters, 236*, Article 111608.

Alessi, L., Di Girolamo, E. F., Pagano, A., & Giudici, M. P. (2024). Accounting for climate transition risk in banks' capital requirements. *Journal of Financial Stability, 73*, Article 101269.

Aouadi, A., & Marsat, S. (2018). Do ESG controversies matter for firm value? Evidence from international data. *Journal of Business Ethics, 151*(4), 1027–1047.

Azmi, W., Hassan, M. K., Houston, R., & Karim, M. S. (2021). ESG activities and banking performance: International evidence from emerging economies. *Journal of International Financial Markets, Institutions and Money, 70*, Article 101277.

Baek, S., & Kang, M. (2025). Does ESG enhance asset quality and funding cost management in banking diversification? *Finance Research Letters, 73*, Article 106542.

Basel Committee on Banking Supervision. (2023). *Disclosure of climate-related financial risks*.

Berger, A. N., Li, X., Saheruddin, H., & Zhao, D. (2024). Government guarantees and bank liquidity creation around the world. *Journal of Banking and Finance, 158*, Article 107048.

Bhagavatula, S., Bhalla, M., Goel, M., & Vissa, B. (2023). Social diversity in corporate boards and firm outcomes. *Journal of Corporate Finance, 83*, Article 102499.

Bigelli, M., Mengoli, S., & Sandri, S. (2023). ESG score, board structure and the impact of the non-financial reporting directive on European firms. *Journal of Economics and Business, 127*, Article 106133.

Birindelli, G., Bonanno, G., Dell'Atti, S., & Iannuzzi, A. P. (2022). Climate change commitment, credit risk and the country's environmental performance: Empirical evidence from a sample of international banks. *Business Strategy and the Environment, 31*(4), 1641–1655.

Boyer, M., & Laffont, J. J. (1997). Environmental risks and bank liability. *European Economic Review, 41*(8), 1427–1459.

Cantero-Saiz, M., Polizzi, S., & Scannella, E. (2024). ESG and asset quality in the banking industry: The moderating role of financial performance. *Research in International Business and Finance, 69*, Article 102221.

Cao, Q., Zhu, T., & Yu, W. (2024). ESG investment and bank efficiency: Evidence from China. *Energy Economics, 133*, Article 107516.

Chalabi-Jabado, F., & Ziane, Y. (2024). Climate risks, financial performance and lending growth: Evidence from the banking industry. *Technological Forecasting and Social Change, 209*, Article 123757.

Chiaramonte, L., Dreassi, A., Girardone, C., & Piserà, S. (2022). Do ESG strategies enhance bank stability during financial turmoil? Evidence from Europe. *The European Journal of Finance, 28*(12), 1173–1211.

Citterio, A., & King, T. (2023). The role of environmental, social, and governance (ESG) in predicting bank financial distress. *Finance Research Letters, 51*, Article 103411.

Coqueret, G., Giroux, T., & Zerbib, O. D. (2025). The biodiversity premium. *Ecological Economics, 228*, Article 108435.

Cosma, S., Cosma, S., Pennetta, D., & Rimo, G. (2025). Does biodiversity matter for firm value? *Journal of International Financial Markets, Institutions and Money, 105*, Article 102240.

Cumming, D., Lu, F., Xu, L., & Yu, C. F. J. (2025). Does stakeholder orientation mitigate shareholder–employee conflicts? Evidence from a quasi-natural experiment. *Journal of Corporate Finance, 91*, Article 102736.

Di Tommaso, C., & Thornton, J. (2020). Do ESG scores effect bank risk taking and value? Evidence from European banks. *Corporate Social Responsibility and Environmental Management, 27*(5), 2286–2298.

Donaldson, T., & Preston, L. E. (1995). The stakeholder theory of the corporation: Concepts, evidence, and implications. *Academy of Management Review, 20*(1), 65–91.

El Khoury, R., Nasrallah, N., & Alareeni, B. (2023). ESG and financial performance of banks in the MENAT region: Concavity–convexity patterns. *Journal of Sustainable Finance and Investment, 13*(1), 406–430.

European Banking Authority. (2024). *Consultation paper Draft Guidelines on the management of ESG risks*.

European Banking Authority. (2025). *Guidelines on the management of environmental, social and governance (ESG) risks*.

Freeman, R. E. (1984). *Strategic management: A stakeholder approach*.

Friedman, M. (2007). The social responsibility of business is to increase its profits. In *Corporate ethics and corporate governance* (pp. 173–178). Springer.

Galletta, S., Goodell, J. W., Mazzù, S., & Paltrinieri, A. (2023). Bank reputation and operational risk: The impact of ESG. *Finance Research Letters, 51*, Article 103494.

García-Meca, E., García-Sánchez, I. M., & Martínez-Ferrero, J. (2015). Board diversity and its effects on bank performance: An international analysis. *Journal of Banking and Finance, 53*, 202–214.

Garel, A., Romec, A., Sautner, Z., & Wagner, A. F. (2024). Do investors care about biodiversity? *Review of Finance, 28*(4), 1151–1186.

Giglio, S., Kuchler, T., Stroebel, J., & Zeng, X. (2025). Biodiversity risk. *Review of Finance, 2025*, rfa063.

Hummel, K., & Jobst, D. (2024). An overview of corporate sustainability reporting legislation in the European Union. *Accounting in Europe, 2024*, 1–36.

Li, S., & Pan, Z. (2022). Climate transition risk and bank performance: Evidence from China. *Journal of Environmental Management, 323*, Article 116275.

Liang, Q., Xu, P., & Jiraporn, P. (2013). Board characteristics and Chinese bank performance. *Journal of Banking and Finance, 37*(8), 2953–2968.

Liu, S., Jin, J., & Nainar, K. (2023). Does ESG performance reduce banks' nonperforming loans? *Finance Research Letters, 55*, Article 103859.

Mulder, I., & Koellner, T. (2011). Hardwiring green: How banks account for biodiversity risks and opportunities. *Journal of Sustainable Finance and Investment, 1*(2), 103–120.

Russo, M. V., & Fouts, P. A. (1997). A resource-based perspective on corporate environmental performance and profitability. *Academy of Management Journal, 40*(3), 534–559.

Scholtens, B., & Dam, L. (2007). Banking on the equator. Are banks that adopted the equator principles different from non-adopters? *World Development, 35*(8), 1307–1328.

Skouloudis, A., Malesios, C., & Dimitrakopoulos, P. G. (2019). Corporate biodiversity accounting and reporting in mega-diverse countries: An examination of indicators disclosed in sustainability reports. *Ecological Indicators, 98*, 888–901.

Smith, N. C., & Rönnegard, D. (2016). Shareholder primacy, corporate social responsibility, and the role of business schools. *Journal of Business Ethics, 134*(3), 463–478.

Tan, B., Igan, D., Peria, M. S. M., Pierri, N., & Presbitero, A. F. (2021). Government intervention and bank markups: Lessons from the global financial crisis for the COVID-19 crisis. *Journal of Banking and Finance, 133*, Article 106320.

Thompson, P., & Cowton, C. J. (2004). Bringing the environment into bank lending: Implications for environmental reporting. *The British Accounting Review, 36*(2), 197–218.

Wu, M. W., & Shen, C. H. (2013). Corporate social responsibility in the banking industry: Motives and financial performance. *Journal of Banking and Finance, 37*(9), 3529–3547.

CHAPTER 7

Conclusions

Abstract This concluding chapter summarises the volume's key findings. Biodiversity loss is a tangible phenomenon, already impacting value chains, credit quality, and the stability of entire sectors, placing banks at the centre of a profound yet still halting transformation. Indeed, biodiversity often remains marginal in risk management and reporting practices, partly due to incomplete metrics, insufficient data, and still limited regulatory guidance. The volume shows that measurement tools are improving and that the financial sector is experimenting with new products, from sustainability-linked loans to conservation bonds, albeit unevenly and with the need for more robust standards. Empirical evidence indicates that banks that integrate environmental and biodiversity risks are more resilient and profitable. The policy message is clear: biodiversity is becoming a central element of financial stability. Supervisors and regulators must provide more precise expectations, credible metrics, and tools to mobilise private capital. Ultimately, the safety of financial systems increasingly depends on the safety of natural systems, and finance must ensure that its activities do not accelerate ecological degradation but contribute, where possible, to nature positive outcomes.

Keywords Nature · Biodiversity loss · Biodiversity finance · Banks · Financial stability

G. Rimo and S. Cosma, *Biodiversity and Banks*, Palgrave Studies in Impact Finance, https://doi.org/10.1007/978-3-032-17631-8_7

7.1 A Brief Recap

The journey explored in this volume started with a simple but troubling observation: the loss of biodiversity is reshaping the foundations of our economies, and the financial system can no longer afford to treat nature as a distant or optional concern. Ecosystem degradation is already having an impact on supply chains, credit quality, productivity, and the stability of entire industries. It is also challenging banks, supervisors, and policymakers to rethink long-standing assumptions about risk, value, and sustainable growth.

Across the chapters, several insights have emerged with remarkable consistency.

The first chapters of this volume have highlighted that biodiversity loss is not something abstract. On the contrary, it is a very concrete phenomenon. We have seen how degraded ecosystems disrupt production, raise default probabilities, weaken collateral values, and amplify volatility across sectors. Firms and banks depend on nature far more deeply than traditional risk models capture. When ecosystems weaken, so does the economic fabric that underpins financial stability.

In this context, banks result in a critical position. They finance activities that can either accelerate ecological decline or support a very different trajectory. At the same time, banks are themselves vulnerable: the evidence reviewed in Chapter 3 shows how biodiversity-related risks can undermine profitability and resilience if left unmanaged. At the same time, the literature review presented in this volume highlights that biodiversity often remains marginal in reporting and risk management practices. The reasons are mostly technical: the lack of standardised metrics, comparable datasets, and regulatory guidance limits the ability of intermediaries to integrate nature-related risks in a structured way.

Despite rapid progress, we still lack a shared language for measuring nature-related risks. Chapter 4 showed that sectoral tools, geographical platforms, and firm-level metrics are all evolving, but none is yet complete. Data are patchy, methods differ, and measurement still struggles to keep pace with the complexity of ecological systems. Yet these tools are improving, and they offer an increasingly valuable foundation for more informed decisions.

In Chapter 5, we show the efforts needed to address the biodiversity finance gap, and highlight that the financial sector is beginning to move, albeit in ways that are still uneven and often fragmented. Banks, in particular, are experimenting with new approaches. From sustainability-linked lending to conservation bonds, from blended finance to biodiversity credits, a growing set of private finance instruments is emerging. Some are promising, others experimental, all are searching for integrity, comparability, and real impact. What is certain is that all these instruments signal the willingness (and possibility) to rethink how capital is allocated and how financial products can support positive transitions for nature.

As discussed in Chapter 5, progress will depend not only on financial innovation but on creating the conditions that allow these instruments to function with integrity. Banks will need reliable standards, guarantees that ensure the credibility of environmental claims, and incentives that make nature-positive lending and investing a rational business choice, not just a voluntary commitment.

Finally, empirical evidence matters. As shown in Chapter 6, banks that integrate biodiversity and environmental risks into their lending processes tend to see benefits in terms of asset quality and profitability, as well as being perceived more positively by the market.

7.2 Implications and Concluding Remarks

Together, our findings suggest a simple but powerful idea, as well as the first implication of this volume: biodiversity is becoming a core element of financial stability and strategic banking behaviour. Banks, more than anyone else, must sense risk. Today risk increasingly comes from unexpected places: from water shortages, degrading soil, uncertain harvests, and commodity markets reacting to environmental shocks. Considering these new risks is not just an exercise in compliance but rather a sign of more robust risk management practices. The institutions that understand this shift early will be better positioned to navigate uncertainty and create long-term value.

The greatest challenges undoubtedly lie with supervisory authorities. Supervisors face a challenge that, until a few years ago, was unimaginable: assessing risks that arise not on balance sheets, but in ecosystems. Banks, as we have seen, can be highly exposed to biodiversity loss, even when they do not operate directly in "environmental" sectors. To address these challenges, supervisory authorities must help banks recognise the risk, providing a clear framework for integrating nature and biodiversity into internal processes, risk committees, and sector policies. Supervision must not only be about control but also about support. Bringing into the discussion issues such as ecologically vulnerable supply chains or collateral exposed to physical risks can improve the quality of the discussion.

For regulators, the main challenge is to guide the financial system in a context where nature-related risks are changing rapidly. Clear and consistent signals are needed on what will be required of banks in the coming years to reduce uncertainty and guide disclosure, due diligence, and risk management practices.

It is equally important to strengthen market integrity: instruments like green bonds, sustainability-linked loans, or biodiversity credits only truly work if based on solid and verifiable metrics.

Finally, regulators must help bridge the enormous nature financing gap, using public levers to mobilise private capital where risks are still too high.

To conclude, the most important implication, and perhaps the simplest: the security of financial systems increasingly depends on the security of natural systems. And that changes everything. It is not the job of banks, regulators, and supervisors to save nature. That is a task for society as a whole. However, banks can ensure that finance does not contribute to the deterioration of nature and, indeed, supports it wherever possible.

Index

G. Rimo and S. Cosma, *Biodiversity and Banks*, Palgrave Studies in Impact Finance, https://doi.org/10.1007/978-3-032-17631-8